Protein-Protein Interactions: Techniques and Applications

Protein-Protein Interactions: Techniques and Applications

Edited by
Brian O'Neill

⊟ Larsen & Keller
www.larsen-keller.com

Protein-Protein Interactions: Techniques and Applications
Edited by Brian O'Neill
ISBN: 978-1-63549-653-6 (Hardback)

© 2018 Larsen & Keller

■ Larsen & Keller

Published by Larsen and Keller Education,
5 Penn Plaza,
19th Floor,
New York, NY 10001, USA

Cataloging-in-Publication Data

Protein-protein interactions : techniques and applications / edited by Brian O'Neill.
    p. cm.
Includes bibliographical references and index.
ISBN 978-1-63549-653-6
1. Protein-protein interactions. 2. Molecular association. I. O'Neill, Brian.
QP551.5 .P76 2018
572.6--dc23

For more information regarding Larsen and Keller Education and its products, please visit the publisher's website www.larsen-keller.com

Table of Contents

Preface

Any interactions between two protein molecules enabled by electrostatic forces like hydrophobic effect is known as protein-protein interaction. The examples of these interactions are cell metabolism, signal transduction, muscle contraction, transport across membranes, etc. The process is used in many subjects namely quantum chemistry, molecular dynamics, biochemistry, etc. This book is a compilation of chapters that discuss the most vital concepts in the field of protein-protein interactions. It discusses in detail the various applications and techniques of the subject. For someone with an interest and eye for detail, this textbook covers the most significant topics in this field. It will serve as a reference to a broad spectrum of readers.

A foreword of all chapters of the book is provided below:

Chapter 1 - Interactome is the entirety of physical interaction that occurs within a cell. It provides fundamental insights into the functions of genomes and proteomes. This chapter is an overview of the subject matter incorporating all the major aspects of protein-protein interactions; **Chapter 2** - Two-hybrid screening is a technique which studies the protein-protein interaction and the protein and DNA interaction. Affinity chromatography, immunoprecipitation, tandem affinity purification and protein mass spectrometry are some of the topics focused upon in this section. The chapter strategically encompasses and incorporates the major components and key concepts of protein-protein interaction, providing a complete understanding; **Chapter 3** - High-throughput method is required when studying large number of proteins simultaneously. Protein microarray is a method to study proteins. Antibody microarrays have also been discussed here. This chapter provides a plethora of interdisciplinary topics for better comprehension of protein-protein interactions; **Chapter 4** - Label-free detection is a detection technique used in protein microarrays. Microcantilever, scanning kelvin nanoprobe, atomic force microscopy, etc. are some label-free detection techniques. This chapter is a compilation of the various branches of label-free detection that form an integral part of the broader subject matter.

At the end, I would like to thank all the people associated with this book devoting their precious time and providing their valuable contributions to this book. I would also like to express my gratitude to my fellow colleagues who encouraged me throughout the process.

Editor

Understanding Interactomics

Interactome is the entirety of physical interaction that occurs within a cell. It provides fundamental insights into the functions of genomes and proteomes. This chapter is an overview of the subject matter incorporating all the major aspects of protein-protein interactions.

Interactomics

Proteins are the most important molecules of all biological entities. They are involved in every aspect of the cell structure and functioning – as backbones for support, cell signaling, hormones, enzymes, mediators of the immune system, and as key players in almost every metabolic pathway of the cell. Despite their wide reach in all the cellular processes, the factor that is arguably more than the protein itself in the cell functioning is the interaction between two proteins. Without the correct interaction between the right protein pairs, no biological function can take place. Conversely, a single incorrect interaction, or aberrant interaction could prove to be disastrous for the cell and possibly for the whole organism. This is one of the most fundamental characteristics in biology, be it for any metabolic pathway in heart, brain, immune system or even at any evolutionary level – right from simple bacteria to plants to humans. Thus, study of protein-protein interactions is one of the most pertinent topics for the understanding of biological functions. Another imperative reason why protein interactions are important to study is because they can cause major difference from normal to diseased conditions. The excess or absence of a single metabolite might set off a chain of downstream processes, which eventually lead to an abnormal, diseased state. Thus, for achieving the crucial goal of understanding the disease, it is necessary to understand cell-signaling pathways involved in it, and how the proteins interact under normal and abnormal conditions.

In molecular biology, an interactome is the whole set of molecular interactions in a particular cell. The term specifically refers to physical interactions among molecules (such as those among proteins, also known as protein–protein interactions, PPIs) but can also describe sets of indirect interactions among genes (genetic interactions). The interactomes based on PPIs should be associated to the proteome of the corresponding species in order to provide a global view ("omic") of all the possible molecular interactions that a protein can present. A recent compendium of interactomes can be obtained in the resource: APID interactomes.

The word "interactome" was originally coined in 1999 by a group of French scientists headed by Bernard Jacq. Mathematically, interactomes are generally displayed as graphs. Though interactomes may be described as biological networks, they should not be confused with other networks such as neural networks or food webs.

Molecular Interaction Networks

Molecular interactions can occur between molecules belonging to different bio-chemical families (proteins, nucleic acids, lipids, carbohydrates, etc.) and also within a given family. Whenever such molecules are connected by physical interactions, they form molecular interaction networks that are generally classified by the nature of the compounds involved. Most commonly, *interactome* refers to *protein–protein interaction* (PPI) network (PIN) or subsets thereof. For instance, the Sirt-1 protein interactome and Sirt family second order interactome is the network involving Sirt-1 and its directly interacting proteins where as second order interactome illustrates interactions up to second order of neighbors (Neighbors of neighbors). Another extensively studied type of interactome is the protein–DNA interactome, also called a *gene-regulatory network*, a network formed by transcription factors, chromatin regulatory proteins, and their target genes. Even *metabolic networks* can be considered as molecular interaction networks: metabolites, i.e. chemical compounds in a cell, are converted into each other by enzymes, which have to bind their substrates physically.

In fact, all interactome types are interconnected. For instance, protein interactomes contain many enzymes which in turn form biochemical networks. Similarly, gene regulatory networks overlap substantially with protein interaction networks and signaling networks.

Size

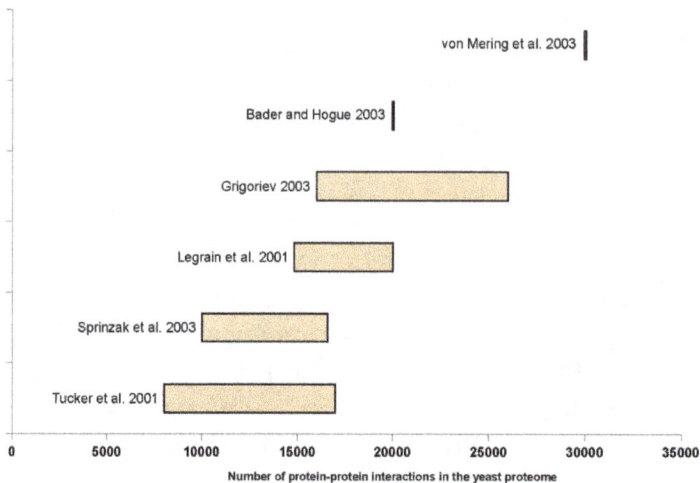

Estimates of the yeast protein interactome

It has been suggested that the size of an organism's interactome correlates better than genome size with the biological complexity of the organism. Although protein–protein interaction maps containing several thousand binary interactions are now available for several species, none of them is presently complete and the size of interactomes is still a matter of debate.

Yeast

The yeast interactome, i.e. all protein–protein interactions among proteins of *Saccharomyces cerevisiae*, has been estimated to contain between 10,000 and 30,000 interactions. A reasonable estimate may be on the order of 20,000 interactions. Larger estimates often include indirect or predicted interactions, often from affinity purification/mass spectrometry (AP/MS) studies.

Genetic Interaction Networks

Genes interact in the sense that they affect each other's function. For instance, a mutation may be harmless, but when it is combined with another mutation, the combination may turn out to be lethal. Such genes are said to "interact genetically". Genes that are connected in such a way form *genetic interaction networks*. Some of the goals of these networks are: develop a functional map of a cell's processes, drug target identification, and to predict the function of uncharacterized genes.

In 2010, the most "complete" gene interactome produced to date was compiled from about 5.4 million two-gene comparisons to describe "the interaction profiles for ~75% of all genes in the budding yeast", with ~170,000 gene interactions. The genes were grouped based on similar function so as to build a functional map of the cell's processes. Using this method the study was able to predict known gene functions better than any other genome-scale data set as well as adding functional information for genes that hadn't been previously described. From this model genetic interactions can be observed at multiple scales which will assist in the study of concepts such as gene conservation. Some of the observations made from this study are that there were twice as many negative as positive interactions, negative interactions were more informative than positive interactions, and genes with more connections were more likely to result in lethality when disrupted.

Interactomics

Interactomics is a discipline at the intersection of bioinformatics and biology that deals with studying both the interactions and the consequences of those interactions between and among proteins, and other molecules within a cell. Interactomics thus aims to compare such networks of interactions (i.e., interactomes) between and within species in order to find how the traits of such networks are either preserved or varied.

Interactomics is an example of "top-down" systems biology, which takes an overhead, as well as overall, view of a biosystem or organism. Large sets of genome-wide and proteomic data are collected, and correlations between different molecules are inferred. From the data new hypotheses are formulated about feedbacks between these molecules. These hypotheses can then be tested by new experiments.

Experimental Methods to Map Interactomes

The study of interactomes is called interactomics. The basic unit of a protein network is the protein–protein interaction (PPI). While there are numerous methods to study PPIs, there are relatively few that have been used on a large scale to map whole interactomes.

The yeast two hybrid system (Y2H) is suited to explore the binary interactions among two proteins at a time. Affinity purification and subsequent mass spectrometry is suited to identify a protein complex. Both methods can be used in a high-throughput (HTP) fashion. Yeast two hybrid screens allow false positive interactions between proteins that are never expressed in the same time and place; affinity capture mass spectrometry does not have this drawback, and is the current gold standard. Yeast two-hybrid data better indicates non-specific tendencies towards sticky interactions rather while affinity capture mass spectrometry better indicates functional in vivo protein–protein interactions.

Computational Methods to Study Interactomes

Once an interactome has been created, there are numerous ways to analyze its properties. However, there are two important goals of such analyses. First, scientists try to elucidate the systems properties of interactomes, e.g. the topology of its interactions. Second, studies may focus on individual proteins and their role in the network. Such analyses are mainly carried out using bioinformatics methods and include the following, among many others:

Validation

First, the coverage and quality of an interactome has to be evaluated. Interactomes are never complete, given the limitations of experimental methods. For instance, it has been estimated that typical Y2H screens detect only 25% or so of all interactions in an interactome. The coverage of an interactome can be assessed by comparing it to benchmarks of well-known interactions that have been found and validated by independent assays. Other methods filter out false positives calculating the similarity of known annotations of the proteins involved or define a likelihood of interaction using the subcellular localization of these proteins.

Predicting PPIs

Using experimental data as a starting point, *homology transfer* is one way to predict interactomes. Here, PPIs from one organism are used to predict interactions among homologous proteins in another organism ("*interologs*"). However, this approach has certain limitations, primarily because the source data may not be reliable (e.g. contain false positives and false negatives). In addition, proteins and their interactions change during evolution and thus may have been lost or gained. Nevertheless, numerous interactomes have been predicted, e.g. that of *Bacillus licheniformis*.

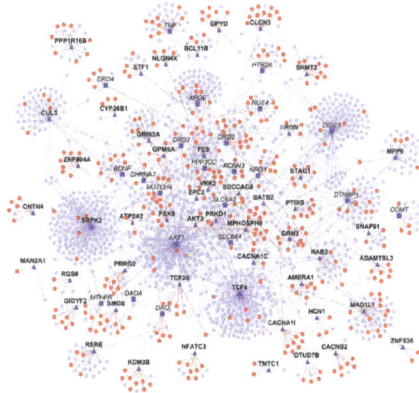

Schziophrenia PPI.

Some algorithms use experimental evidence on structural complexes, the atomic details of binding interfaces and produce detailed atomic models of protein–protein complexes as well as other protein–molecule interactions. Other algorithms use only sequence information, thereby creating unbiased complete networks of interaction with many mistakes.

Some methods use machine learning to distinguish how interacting protein pairs differ from non-interacting protein pairs in terms of pairwise features such as cellular colocalization, gene co-expression, how closely located on a DNA are the genes that encode the two proteins, and so on.Random Forest has been found to be most-effective machine learning method for protein interaction prediction. Such methods have been applied for discovering protein interactions on human interactome, specifically the interactome of Membrane proteins and the interactome of Schizophrenia-associated proteins.

Text Mining of PPIs

Some efforts have been made to extract systematically interaction networks directly from the scientific literature. Such approaches range in terms of complexity from simple co-occurrence statistics of entities that are mentioned together in the same context (e.g. sentence) to sophisticated natural language processing and machine learning methods for detecting interaction relationships.

Protein Function Prediction

Protein interaction networks have been used to predict the function of proteins of unknown functions. This is usually based on the assumption that uncharacterized proteins have similar functions as their interacting proteins (*guilt by association*). For example, YbeB, a protein of unknown function was found to interact with ribosomal proteins and later shown to be involved in translation. Although such predictions may be based on single interactions, usually several interactions are found. Thus, the whole network of interactions can be used to predict protein functions, given that certain functions are usually enriched among the interactors.

Perturbations and Disease

The *topology* of an interactome makes certain predictions how a network reacts to the perturbation (e.g. removal) of nodes (proteins) or edges (interactions). Such perturbations can be caused by mutations of genes, and thus their proteins, and a network reaction can manifest as a disease. A network analysis can identified drug targets and biomarkers of diseases.

Network Structure and Topology

Interaction networks can be analyzed using the tools of graph theory. Network properties include the degree distribution, clustering coefficients, betweenness centrality, and many others. The distribution of properties among the proteins of an interactome has revealed that the interactome networks often have scale-free topology where functional modules within a network indicate specialized subnetworks. Such modules can be functional, as in a signaling pathway, or structural, as in a protein complex. In fact, it is a formidable task to identify protein complexes in an interactome, given that a network on its own does not directly reveal the presence of a stable complex.

Studied Interactomes

Viral Interactomes

Viral protein interactomes consist of interactions among viral or phage proteins. They were among the first interactome projects as their genomes are small and all proteins can be analyzed with limited resources. Viral interactomes are connected to their host interactomes, forming virus-host interaction networks. Some published virus interactomes include

Bacteriophage

- *Escherichia coli* bacteriophage lambda
- *Escherichia coli* bacteriophage T7
- *Streptococcus pneumoniae* bacteriophage Dp-1
- *Streptococcus pneumoniae* bacteriophage Cp-1

The lambda and VZV interactomes are not only relevant for the biology of these viruses but also for technical reasons: they were the first interactomes that were mapped with multiple Y2H vectors, proving an improved strategy to investigate interactomes more completely than previous attempts have shown.

Human (Mammalian) Viruses

- Human varicella zoster virus (VZV)

- Chandipura virus

- Epstein-Barr virus (EBV)

- Hepatitis C virus (HPC), Human-HCV interactions

- Hepatitis E virus (HEV)

- Herpes simplex virus 1 (HSV-1)

- Kaposi's sarcoma-associated herpesvirus (KSHV)

- Murine cytomegalovirus (mCMV)

Bacterial Interactomes

Relatively few bacteria have been comprehensively studied for their protein–protein interactions. However, none of these interactomes are complete in the sense that they captured all interactions. In fact, it has been estimated that none of them covers more than 20% or 30% of all interactions, primarily because most of these studies have only employed a single method, all of which discover only a subset of interactions. Among the published bacterial interactomes (including partial ones) are

Species	Proteins Total	Interactions	Type
Helicobacter pylori	1,553	~3,004	Y2H
Campylobacter jejuni	1,623	11,687	Y2H
Treponema pallidum	1,040	3,649	Y2H
Escherichia coli	4,288	(5,993)	AP/MS
Escherichia coli	4,288	2,234	Y2H
Mesorhizobium loti	6,752	3,121	Y2H
Mycobacterium tuberculosis	3,959	>8000	B2H
Mycoplasma genitalium	482		AP/MS
Synechocystis sp. PCC6803	3,264	3,236	Y2H
Staphylococcus aureus (MRSA)	2,656	13,219	AP/MS

The *E. coli* and *Mycoplasma* interactomes have been analyzed using large-scale protein complex affinity purification and mass spectrometry (AP/MS), hence it is not easily possible to infer direct interactions. The others have used extensive yeast two-hybrid (Y2H) screens. The *Mycobacterium tuberculosis* interactome has been analyzed using a bacterial two-hybrid screen (B2H).

Eukaryotic Interactomes

There have been several efforts to map eukaryotic interactomes through HTP methods. While no biological interactomes have been fully characterized, over 90% of proteins in *Saccharomyces cerevisiae* have been screened and their interactions characterized, making it the best-characterized interactome. Species whose interactomes have been studied in some detail include:

- *Schizosaccharomyces pombe*

- *Caenorhabditis elegans*

- *Drosophila melanogaster*

- *Homo sapiens*

Recently, the pathogen-host interactomes of Hepatitis C Virus/Human (2008), Epstein Barr virus/Human (2008), Influenza virus/Human (2009) were delineated through HTP to identify essential molecular components for pathogens and for their host's immune system.

Predicted Interactomes

As described above, PPIs and thus whole interactomes can be predicted. While the reliability of these predictions is debatable, they are providing hypotheses that can be tested experimentally. Interactomes have been predicted for a number of species, e.g.

- Human (*Homo sapiens*)

- Rice (*Oryza sativa*)

- *Xanthomonas oryzae*

- *Arabidopsis thaliana*

- Tomato

- *Brassica rapa*

- Maize, corn (*Zea mays*)

- *Populus trichocarpa*

Evolution

The evolution of interactome complexity is delineated in a study published in *Nature*. In this study it is first noted that the boundaries between prokaryotes, unicellular eukaryotes and multicellular eukaryotes are accompanied by orders-of-magnitude reductions in effective population size, with concurrent amplifications of the effects of random genetic drift. The resultant decline in the efficiency of selection seems to be sufficient to influence a wide range of attributes at the genomic level in a nonadaptive manner. The Nature study shows that the variation in the power of random genetic drift is also capable of influencing phylogenetic diversity at the subcellular and cellular levels. Thus, population size would have to be considered as a potential determinant of the mechanistic pathways underlying long-term phenotypic evolution. In the study it is further shown that a phylogenetically broad inverse relation exists between the power of drift and the structural integrity of protein subunits. Thus, the accumulation of mildly deleterious mutations in populations of small size induces secondary selection for protein–protein interactions that stabilize key gene functions, mitigating the structural degradation promoted by inefficient selection. By this means, the complex protein architectures and interactions essential to the genesis of phenotypic diversity may initially emerge by non-adaptive mechanisms.

Criticisms, Challenges, and Responses

Kiemer and Cesareni raise the following concerns with the state (circa 2007) of the field especially with the comparative interactomic: The experimental procedures associated with the field are error prone leading to "noisy results". This leads to 30% of all reported interactions being artifacts. In fact, two groups using the same techniques on the same organism found less than 30% interactions in common. However, some authors have argued that such non-reproducibility results from the extraordinary sensitivity of various methods to small experimental variation. For instance, identical conditions in Y2H assays result in very different interactions when different Y2H vectors are used.

Techniques may be biased, i.e. the technique determines which interactions are found. In fact, any method has built in biases, especially protein methods. Because every protein is different no method can capture the properties of each protein. For instance, most analytical methods that work fine with soluble proteins deal poorly with membrane proteins. This is also true for Y2H and AP/MS technologies.

Interactomes are not nearly complete with perhaps the exception of *S. cerevisiae*. This is not really a criticism as any scientific area is "incomplete" initially until the methodologies have been improved. Interactomics in 2015 is where genome sequencing was in the late 1990s, given that only a few interactome datasets are available.

While genomes are stable, interactomes may vary between tissues, cell types, and developmental stages. Again, this is not a criticism, but rather a description of the challenges in the field.

It is difficult to match evolutionarily related proteins in distantly related species. While homologous DNA sequences can be found relatively easily, it is much more difficult to predict homologous interactions ("interologs") because the homologs of two interacting proteins do not need to interact. For instance, even within a proteome two proteins may interact but their paralogs may not.

Each protein–protein interactome may represent only a partial sample of potential interactions, even when a supposedly definitive version is published in a scientific journal. Additional factors may have roles in protein interactions that have yet to be incorporated in interactomes. The binding strength of the various protein interactors, microenvironmental factors, sensitivity to various procedures, and the physiological state of the cell all impact protein–protein interactions, yet are usually not accounted for in interactome studies.

Biological Network

A biological network is any network that applies to biological systems. A network is any system with sub-units that are linked into a whole, such as species units linked into a whole food web. Biological networks provide a mathematical representation of connections found in ecological, evolutionary, and physiological studies, such as neural networks. The analysis of biological networks with respect to human diseases has led to the field of network medicine.

Network Biology and Bioinformatics

Complex biological systems may be represented and analyzed as computable networks. For example, ecosystems can be modeled as networks of interacting species or a protein can be modeled as a network of amino acids. Breaking a protein down farther, amino acids can be represented as a network of connected atoms, such as carbon, nitrogen, and oxygen. Nodes and edges are the basic components of a network. Nodes represent units in the network, while edges represent the interactions between the units. Nodes can represent a wide-array of biological units, from individual organisms to individual neurons in the brain. Two important properties of a network are degree and betweenness centrality. Degree (or connectivity, a distinct usage from that used in graph theory) is the number of edges that connect a node, while betweenness is a measure of how central a node is in a network. Nodes with high betweenness essentially serve as bridges between different portions of the network (i.e. interactions must pass through this node to reach other portions of the network). In social networks, nodes with high degree or high betweenness may play important roles in the overall composition of a network.

As early as the 1980s, researchers started viewing DNA or genomes as the dynamic storage of a language system with precise computable finite states represented as a finite state machine. Recent complex systems research has also suggested some far-reaching commonality in the organization of information in problems from biology, computer science, and physics, such as the Bose–Einstein condensate (a special state of matter).

Bioinformatics has increasingly shifted its focus from individual genes, proteins, and search algorithms to large-scale networks often denoted as -omes such as biome, interactome, genome and proteome. Such theoretical studies have revealed that biological networks share many features with other networks such as the Internet or social networks, e.g. their network topology.

Networks in Biology

Protein–protein Interaction Networks

Many protein–protein interactions (PPIs) in a cell form *protein interaction networks* (PINs) where proteins are *nodes* and their interactions are *edges*. PINs are the most intensely analyzed networks in biology. There are dozens of PPI detection methods to identify such interactions. The yeast two-hybrid system is a commonly used experimental technique for the study of binary interactions.

Recent studies have indicated conservation of molecular networks through deep evolutionary time. Moreover, it has been discovered that proteins with high degrees of connectedness are more likely to be essential for survival than proteins with lesser degrees. This suggests that the overall composition of the network (not simply interactions between protein pairs) is important for the overall functioning of an organism.

Gene Regulatory Networks (DNA–protein Interaction Networks)

The activity of genes is regulated by transcription factors, proteins that typically bind to DNA. Most transcription factors bind to multiple binding sites in a genome. As a result, all cells have complex gene regulatory networks. For instance, the human genome encodes on the order of 1,400 DNA-binding transcription factors that regulate the expression of more than 20,000 human genes. Technologies to study gene regulatory networks include ChIP-chip, ChIP-seq, CliP-seq, and others.

Gene Co-expression Networks (Transcript–transcript Association Networks)

Gene co-expression networks can be interpreted as association networks between variables that measure transcript abundances. These networks have been used to provide a systems biologic analysis of DNA microarray data, RNA-seq data, miRNA data etc. weighted gene co-expression network analysis is widely used to identify co-expression modules and intramodular hub genes. Co-expression modules may correspond to cell types or pathways. Highly connected intramodular hubs can be interpreted as representatives of their respective module.

Metabolic Networks

The chemical compounds of a living cell are connected by biochemical reactions which convert one compound into another. The reactions are catalyzed by enzymes. Thus, all compounds in a cell are parts of an intricate biochemical network of reactions which is called metabolic network. It is possible to use network analyses to infer how selection acts on metabolic pathways.

Signaling Networks

Signals are transduced within cells or in between cells and thus form complex signaling networks. For instance, in the MAPK/ERK pathway is transduced from the cell surface to the cell nucleus by a series of protein–protein interactions, phosphorylation reactions, and other events. Signaling networks typically integrate protein–protein interaction networks, gene regulatory networks, and metabolic networks.

Neuronal Networks

The complex interactions in the brain make it a perfect candidate to apply network theory. Neurons in the brain are deeply connected with one another and this results in complex networks being present in the structural and functional aspects of the brain. For instance, small-world network properties have been demonstrated in connections between cortical areas of the primate brain. This suggests that cortical areas of the brain are not directly interacting with each other, but most areas can be reached from all others through only a few interactions.

Food Webs

All organisms are connected to each other through feeding interactions. That is, if a species eats or is eaten by another species, they are connected in an intricate food web of predator and prey interactions. The stability of these interactions has been a long-standing question in ecology. That is to say, if certain individuals are removed, what happens to the network (i.e. does it collapse or adapt)? Network analysis can be used to explore food web stability and determine if certain network properties result in more stable networks. Moreover, network analysis can be used to determine how selective removals of species will influence the food web as a whole. This is especially important considering the potential species loss due to global climate change.

Between-species Interaction Networks

In biology, pairwise interactions have historically been the focus of intense study. With the recent advances in network science, it has become possible to scale up pairwise interactions to include individuals of many species involved in many sets of interactions to understand the structure and function of larger ecological networks. The use of network analysis can allow for both the discovery and understanding how these complex interactions link together within the system's network, a property which has previously been overlooked. This powerful tool allows for the study of various types of interactions (from competitive to cooperative) using the same general framework. For example, plant-pollinator interactions are mutually beneficial and often involve many different species of pollinators as well as many different species of plants. These interactions are critical to plant reproduction and thus the accumulation of resources at the base of the food chain for primary consumers, yet these interaction networks are threatened by anthropogenic change. The use of network analysis can illuminate how pollination networks work and may in turn inform conservation efforts. Within pollination networks, nestedness (i.e., specialists interact with a subset of species that generalists interact with), redundancy (i.e., most plants are pollinated by many pollinators), and modularity play a large role in network stability. These network properties may actually work to slow the spread of disturbance effects through the system and potentially buffer the pollination network from anthropogenic changes somewhat. More generally, the structure of species interactions within an ecological network can tell us something about the diversity, richness, and robustness of the network. Researchers can even compare current constructions of species interactions networks with historical reconstructions of ancient networks to determine how networks have changed over time. Recent research into these complex species interactions networks is highly concerned with understanding what factors (e.g., diversity) lead to network stability.

Within-species Interaction Networks

Network analysis provides the ability to quantify associations between individuals, which makes it possible to infer details about the network as a whole at the species and/or population level. Researchers interested in animal behavior across a multitude of taxa, from insects to primates, are starting to incorporate network analysis into their research. Researchers interested in social insects (e.g., ants and bees) have used network analyses to better understand division of labor, task allocation, and foraging optimization within colonies; Other researchers are interested in how certain network properties at the group and/or population level can explain individual level behaviors. For instance, a study on wire-tailed manakins (a small passerine bird) found that a male's degree in the network largely predicted the ability of the male to rise in the social hierarchy (i.e. eventually obtain a territory and matings). In bottlenose dolphin groups, an individual's degree and betweenness centrality values may predict whether or not that individual will exhibit certain behaviors, like the use of side flopping and upside-down lobtailing to lead group traveling efforts; individuals with high betweenness values are more connected and can obtain more information, and thus are better suited to lead group travel and therefore tend to exhibit these signaling behaviors more than other group members. Network analysis can also be used to describe the social organization within a species more generally, which frequently reveals important proximate mechanisms promoting the use of certain behavioral strategies. These descriptions are frequently linked to ecological properties (e.g., resource distribution).

For example, network analyses revealed subtle differences in the group dynamics of two related equid fission-fusion species, Grevy's zebra and onagers, living in variable environments; Grevy's zebras show distinct preferences in their association choices when they fission into smaller groups, whereas onagers do not. Similarly, researchers interested in primates have also utilized network analyses to compare social organizations across the diverse primate order, suggesting that using network measures (such as centrality, assortativity, modularity, and betweenness) may be useful in terms of explaining the types of social behaviors we see within certain groups and not others. Finally, social network analysis can also reveal important fluctuations in animal behaviors across changing environments. For example, network analyses in female chacma baboons (*Papio hamadryas ursinus*) revealed important dynamic changes across seasons which were previously unknown; instead of creating stable, long-lasting social bonds with friends, baboons were found to exhibit more variable relationships which were dependent on short-term contingencies related to group level dynamics as well as environmental variability. This is a very small set of broad examples of how researchers can use network analysis to study animal behavior. Research in this area is currently expanding very rapidly. Social network analysis is a valuable tool for studying animal behavior across all animal species, and has the potential to uncover new information about animal behavior and social ecology that was previously poorly understood.

Significance of Interactomics

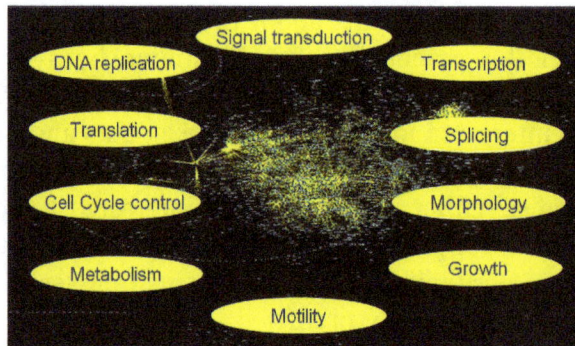

Interactomics is the study of a network of interactions. Interactome consists of all the interactions among biological pathways and the associated molecules. Interactomics utilizes bioinformatics approach along with the experimental data. The necessity for the usage of bioinformatics can be attributed to the immense amount of information obtained for each network – the key players, their upstream/downstream interactors, the type of interaction etc. But in all the interactions the most active players are the proteins – almost all major steps in all the pathways are mediated by proteins in the form of enzymes, hormones, receptors, metabolites etc. Therefore, to understand the mechanism of cellular molecular processes, studying the protein-protein interaction becomes most essential.

Protein-protein interactions: the potential cellular processes, which may get affected due to improper protein-protein interactions. There are different ways in which these interactions take place. These interactions can be strong or weak, or they may be transient or permanent and may finally result in formation of homo-oligomers or hetero-oligomers. Any alteration in these interactions could lead to formation of new, incorrect oligomers or render a key factor inactive, or a change in the kinetics of a certain reaction.

Each interactome is highly controlled and tightly regulated. Any imbalance can hamper normal conditions and result in a disease-like state. Thus, studying the protein interactions is important as they interact with a wide variety of biomolecules such as lipids, nucleic acids, small drug inhibitors, etc. Proteins also interact with one another to form macromolecular complexes that regulate signal transduction & gene regulation. Study of these networks could also help in understanding the function of uncharacterized proteins, and also to find out new roles for characterized proteins. Also, new networks of protein interactions can be found out as the same protein may play a role in different pathways. This understanding will further help us to manipulate the networks, thereby opening new avenues to study the diseases.

Protein–protein Interaction

Proteins are responsible for several functions in a cell ranging from a catalyzing reaction to several complex functions. Protein- protein interaction plays an important role within the cellular machinery for specific functions like signal transduction, translation, transcription, replication, control on Cell Cycle etc. The interaction between protein molecules leads to the formation of a larger protein complexes performing a specific function. Interaction among proteins depend upon various factors associated with the protein itself, such as its amino acid sequence, associated co-factors, and finally most important is the three-dimensional shape of protein. The different forces which help in the protein- protein interaction include various non-covalent interaction such as Van-der Waals force, hydrogen bonds formation etc.

Protein–protein interactions (PPIs) are the physical contacts of high specificity established between two or more protein molecules as a result of biochemical events steered by electrostatic forces including the hydrophobic effect. Many are physical contacts with molecular associations between chains that occur in a cell or in a living organism in a specific biomolecular context.

Proteins rarely act alone as their functions tend to be regulated. Many molecular processes within a cell are carried out by molecular machines that are built from a large number of protein components organized by their PPIs. These interactions make up the so-called interactomics of the organism, while aberrant PPIs are the basis of multiple aggregation-related diseases, such as Creutzfeldt–Jakob, Alzheimer's disease, and may lead to cancer.

PPIs have been studied from different perspectives: biochemistry, quantum chemistry, molecular dynamics, signal transduction, among others. All this information enables the creation of large protein interaction networks – similar to metabolic or genetic/epigenetic networks – that empower the current knowledge on biochemical cascades and molecular etiology of disease, as well as the discovery of putative protein targets of therapeutic interest.

Examples

Signal Transduction

The activity of the cell is regulated by extracellular signals. Signals propagation to inside and/or along the interior of cells depends on PPIs between the various signaling molecules. The recruitment of signaling pathways through PPIs is called signal transduction and plays a fundamental role in many biological processes and in many diseases including Parkinson's disease and cancer.

Transport Across Membranes

A protein may be carrying another protein (for example, from cytoplasm to nucleus or vice versa in the case of the nuclear pore importins).

Cell Metabolism

In many biosynthetic processes enzymes interact with each other to produce small compounds or other macromolecules.

Muscle Contraction

Physiology of muscle contraction involves several interactions. Myosin filaments act as molecular motors and by binding to actin enables filament sliding. Furthermore, members of the skeletal musclelipid droplet-associated proteins family associate with other proteins, as activator of adipose triglyceride lipase and its coactivator comparative gene identification-58, to regulate lipolysis in skeletal muscle.

Types

To describe the types of protein–protein interactions (PPIs) it is important to consider that proteins can interact in a "transient" way (to produce some specific effect in a short time) or to interact with other proteins in a "stable" way to build multiprotein complexes that are molecular machines within the living systems. A protein complex assembly can result in the formation of homo-oligomeric or hetero-oligomeric complexes. In addition to the conventional complexes, as enzyme-inhibitor and antibody-antigen, interactions can also be established between domain-domain and domain-peptide. Another important distinction to identify protein-protein interactions is the way they have been determined, since there are techniques that measure direct physical interactions between protein pairs, named "binary" methods, while there are other techniques that measure physical interactions among groups of proteins, without pairwise determination of protein partners, named "co-complex" methods.

Homo-oligomers vs. Hetero-oligomers

Homo-oligomers are macromolecular complexes constituted by only one type of protein subunit. Protein subunits assembly is guided by the establishment of non-covalent interactions in the quaternary structure of the protein. Disruption of homo-oligomers in order to return to the initial individual monomers often requires denaturation of the complex. Several enzymes, carrier proteins, scaffolding proteins, and transcriptional regulatory factors carry out their functions as homo-oligomers. Distinct protein subunits interact in hetero-oligomers, which are essential to control several cellular functions. The importance of the communication between heterologous proteins is even more evident during cell signaling events and such interactions are only possible due to structural domains within the proteins.

Stable Interactions vs. Transient Interactions

Stable interactions involve proteins that interact for a long time, taking part of permanent complexes as subunits, in order to carry out structural or functional roles. These are usually the case of homo-oligomers (e.g. cytochrome c), and some hetero-oligomeric proteins, as the subunits of ATPase. On the other hand, a protein may interact briefly and in a reversible manner with other proteins in only certain cellular contexts – cell type, cell cycle stage, external factors, presence of other binding proteins, etc. – as it happens with most of the proteins involved in biochemical cascades. These are called transient interactions. For example, some G protein-coupled receptors only transiently bind to $G_{i/o}$ proteins when they are activated by extracellular ligands, while some G_q-coupled receptors, such as muscarinic receptor M3, pre-couple with G_q proteins prior to the receptor-ligand binding. Interactions between intrinsically disordered protein regions to globular protein domains (i.e. MoRFs) are transient interactions.

Covalent vs. non-covalent

Covalent interactions are those with the strongest association and are formed by disulphide bonds or electron sharing. Although being rare, these interactions are determinant in some posttranslational modifications, as ubiquitination and SUMOylation. Non-covalent bonds are usually established during transient interactions by the combination of weaker bonds, such as hydrogen bonds, ionic interactions, Van der Waals forces, or hydrophobic bonds.

Role of Water

Water molecules play a significant role in the interactions between proteins. The crystal structures of complexes, obtained at high resolution from different but homologous proteins, have shown that some interface water molecules are conserved between homologous complexes. The majority of the interface water molecules make hydrogen bonds with both partners of each complex. Some interface amino acid residues or atomic groups of one protein partner engage in both direct and water mediated interactions with the other protein partner. Doubly indirect interactions, mediated by two water molecules, are more numerous in the homologous complexes of low affinity. Carefully conducted mutagenesis experiments, e.g. changing a tyrosine residue into a phenylalanine, have shown that water mediated interactions can contribute to the energy of interaction. Thus, water molecules may facilitate the interactions and cross-recognitions between proteins.

Structure

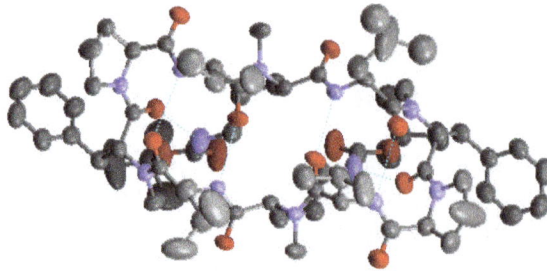

Crystal structure of modified Gramicidin S horizontally determined by X-ray crystallography

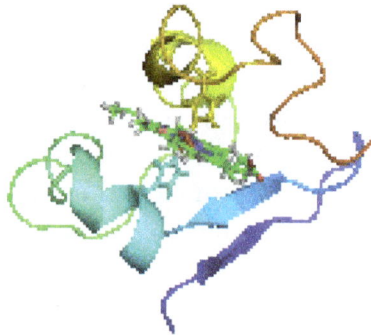

NMR structure of cytochrome C illustrating its dynamics in solution

The molecular structures of many protein complexes have been unlocked by the technique of X-ray crystallography. The first structure to be solved by this method was that of sperm whalemyoglobin by Sir John Cowdery Kendrew. In this technique the angles and intensities of a beam of X-rays diffracted by crystalline atoms are detected in a film, thus producing a three-dimensional picture of the density of electrons within the crystal.

Later, nuclear magnetic resonance also started to be applied with the aim of unravelling the molecular structure of protein complexes. One of the first examples was the structure of calmodulin-binding domains bound to calmodulin. This technique is based on the study of magnetic properties of atomic nuclei, thus determining physical and chemical properties of the correspondent atoms or the molecules. Nuclear magnetic resonance is advantageous for characterizing weak PPIs.

Domains

Proteins hold structural domains that allow their interaction with and bind to specific sequences on other proteins:

- Src homology 2 (SH2) domain

  SH2 domains are structurally composed by three-stranded twisted beta sheet sandwiched flanked by two alpha-helices. The existence of a deep binding pocket with high affinity for phosphotyrosine, but not for phosphoserine or phosphothreonine, is essential for the recognition of tyrosine phosphorylated proteins, mainly autophosphorylated growth factor receptors. Growth factor receptor binding proteins and phospholipase Cγ are examples of proteins that have SH2 domains.

- Src homology 3 (SH3) domain

  Structurally, SH3 domains are constituted by a beta barrel formed by two orthogonal beta sheets and three anti-parallel beta strands. These domains recognize proline enriched sequences, as polyproline type II helical structure (PXXP motifs) in cell signaling proteins like protein tyrosine kinases and the growth factor receptor bound protein 2 (Grb2).

- Phosphotyrosine-binding (PTB) domain

  PTB domains interact with sequences that contain a phosphotyrosine group. These domains can be found in the insulin receptor substrate.

- LIM domain

  LIM domains were initially identified in three homeodomain transcription factors (lin11, is11, and mec3). In addition to this homeodomain proteins and other proteins involved in development, LIM domains have also been identified in non-homeodomain proteins with relevant roles in cellular differentiation, association with cytoskeleton and senescence. These domains contain a tandem cysteine-rich Zn^{2+}-finger motif and embrace the consensus sequence $CX2CX16-23HX2CX2CX2CX16-21CX2C/H/D$. LIM domains bind to PDZ domains, bHLH transcription factors, and other LIM domains.

- *Sterile alpha motif (SAM) domain*

  SAM domains are composed by five helices forming a compact package with a conserved hydrophobic core. These domains, which can be found in the Eph receptor and the stromal interaction molecule (STIM) for example, bind to non-SAM domain-containing proteins and they also appear to have the ability to bind RNA.

- *PDZ domain*

  PDZ domains were first identified in three guanylate kinases: PSD-95, DlgA and ZO-1. These domains recognize carboxy-terminal tri-peptide motifs (S/TXV), other PDZ domains or LIM domains and bind them through a short peptide sequence that has a C-terminal hydrophobic residue. Some of the proteins identified as having PDZ domains are scaffolding proteins or seem to be involved in ion receptor assembling and receptor-enzyme complexes formation.

- *FERM domain*

  FERM domains contain basic residues capable of binding $PtdIns(4,5)P_2$. Talin and focal adhesion kinase (FAK) are two of the proteins that present FERM domains.

- *Calponin homology (CH) domain*

  CH domains are mainly present in cytoskeletal proteins as parvin.

- *Pleckstrin homology domain*

  Pleckstrin homology domains bind to phosphoinositides and acid domains in signaling proteins.

- *WW domain*

  WW domains bind to proline enriched sequences.

- *WSxWS motif*

  Found in cytokine receptors

Properties of the Interface

The study of the molecular structure can give fine details about the interface that enables the interaction between proteins. When characterizing PPI interfaces it is important to take into account the type of complex.

Parameters evaluated include size (measured in absolute dimensions $Å^2$ or in solvent-accessible surface area (SASA)), shape, complementarity between surfaces, residue interface propensities, hydrophobicity, segmentation and secondary structure, and conformational changes on complex formation.

The great majority of PPI interfaces reflects the composition of protein surfaces, rather than the protein cores, in spite of being frequently enriched in hydrophobic residues, particularly in aromatic residues. PPI interfaces are dynamic and frequently planar, although they can be globular and protruding as well. Based on three structures – insulin dimer, trypsin-pancreatic trypsin inhibitor complex, and oxyhaemoglobin – Cyrus Chothia and Joel Janin found that between 1,130 and 1,720 Å² of surface area was removed from contact with water indicating that hydrophobicity is a major factor of stabilization of PPIs. Later studies refined the buried surface area of the majority of interactions to 1,600±350 Å². However, much larger interaction interfaces were also observed and were associated with significant changes in conformation of one of the interaction partners. PPIs interfaces exhibit both shape and electrostatic complementarity.

Regulation

- Protein concentration, which in turn are affected by expression levels and degradation rates;

- Protein affinity for proteins or other binding ligands;

- Ligands concentrations (substrates, ions, etc.);

- Presence of other proteins, nucleic acids, and ions;

- Electric fields around proteins;

- Occurrence of covalent modifications;

Measurement

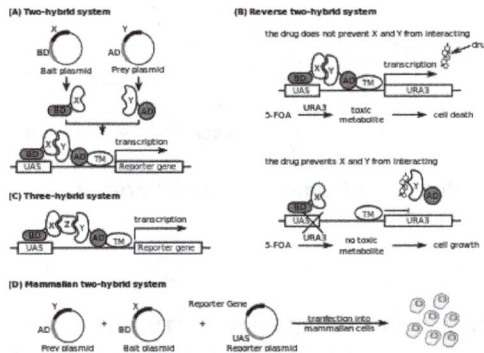

Principles of yeast and mammalian two-hybrid systems

There are a multitude of methods to detect them. Each of the approaches has its own strengths and weaknesses, especially with regard to the sensitivity and specificity of the method. The most conventional and widely used high-throughput methods are yeast two-hybrid screening and affinity purification coupled to mass spectrometry.

Yeast Two-hybrid Screening

This system was firstly described in 1989 by Fields and Song using *Saccharomyces cerevisiae* as biological model. Yeast two hybrid allows the identification of pairwise PPIs (binary method) *in vivo*, indicating non-specific tendencies towards sticky interactions.

Yeast cells are transfected with two plasmids: the bait (protein of interest fused with the DNA-binding domain of a yeast transcription factor, like Gal4), and the prey (a library of cDNA fragments linked to the activation domain of the transcription factor. Transcription of reporter genes does not occur unless bait and prey interact with each other and form a functional transcription factor. Thus, the interaction between proteins can be inferred by the presence of the products resultant of the reporter gene expression.

Despite its usefulness, the yeast two-hybrid system has limitations: specificity is relatively low; uses yeast as main host system, which can be a problem when studying other biological models; the number of PPIs identified is usually low because some transient PPIs are lost during purification steps; and, understates membrane proteins, for example. Limitations have been overcoming by the emergence of yeast two-hybrid variants, such as the membrane yeast two-hybrid (MYTH) and the split-ubiquitin system, which are not limited to interactions that occur in the nucleus; and, the bacterial two-hybrid system, performed in bacteria.

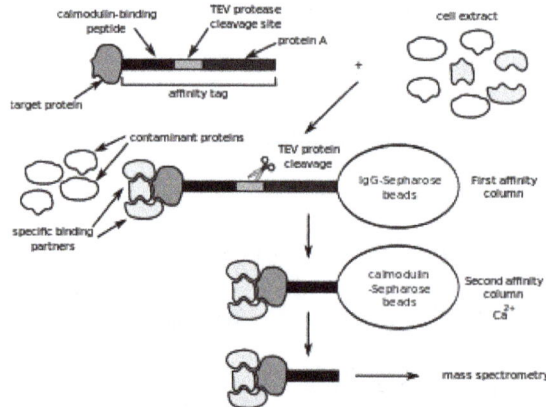

Principle of Tandem Affinity Purification

Affinity Purification Coupled to Mass Spectrometry

Affinity purification coupled to mass spectrometry mostly detects stable interactions and thus better indicates functional in vivo PPIs. This method starts by purification of the tagged protein, which is expressed in the cell usually at *in vivo* concentrations, and its interacting proteins (affinity purification). One of the most advantageous and widely used method to purify proteins with very low contaminating background is the tandem affinity purification, developed by Bertrand Seraphin and Mathias Mann and respective colleagues. PPIs can then be quantitatively and qualitatively analysed by mass spectrometry using different methods: chemical incorporation, biological or metabolic incorporation (SILAC), and label-free methods.

Other Potential Methods

Diverse techniques to identify PPIs have been emerging along with technology progression. These include co-immunoprecipitation, protein microarrays, analytical ultracentrifugation, light scattering, fluorescence spectroscopy, luminescence-based mammalian interactome mapping (LUMIER), resonance-energy transfer systems, mammalian protein–protein interaction trap, electro-switchable biosurfaces, protein-fragment complementation assay, as well as real-time label-free measurements by surface plasmon resonance, and calorimetry.

Text Mining Methods

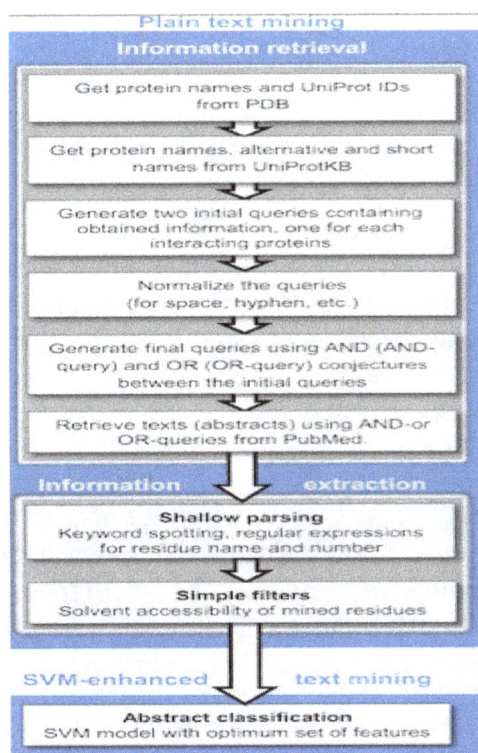

Text mining protocol.

Publicly available information from biomedical research is readily accessible through the internet and is becoming a powerful resource for predictive protein-protein interactions and protein docking. Text mining is much less time costly and consuming compared to other high-throughput techniques. Currently, these methods generally detect binary relations between interacting protein from individual sentences using machine learning and rule/pattern-based information extraction and machine learning approaches. A wide variety of text mining predicting PPIs applications are available for public use, as well as repositories which often stores manually validated and/or computationally predicted PPIs. The principal stages of text mining divides the implementation into two stages: *information retrieval,* where literature abstracts containing names of either or both proteins complexes are selected and *information extraction,* where detecting occurrences of residues are retrieved. The extraction is automated by searching for co-existing sentences, abstracts or paragraphs within textual context.

There are also studies using phylogenetic profiling, basing their functionalities on the theory that proteins involved in common pathways co-evolve in a correlated fashion across large number of species. More complex text mining methodologies use advanced dictionaries and generate networks by Natural Language Processing (NLP) of text, considering gene names as nodes and verbs as edges, other developments involve kernel methods to predict protein interactions.

Machine Learning Methods

These methods use machine learning to distinguish how interacting protein pairs differ from non-interacting protein pairs in terms of pairwise features such as cellular colocalization, gene co-expression, how closely located on a DNA are the genes that encode the two proteins, and so on.Random Forest has been found to be most-effective machine learning method for protein interaction prediction. Such methods have been applied for discovering protein interactions on human interactome, specifically the interactome of Membrane proteins and the interactome of Schizophrenia-associated proteins.

Databases

Large scale identification of PPIs generated hundreds of thousands interactions, which were collected together in specialized biological databases that are continuously updated in order to provide complete interactomes. The first of these databases was the Database of Interacting Proteins (DIP). Since that time, the number of public databases has been increasing. Databases can be subdivided into primary databases, meta-databases, and prediction databases.

Primary databases collect information about published PPIs proven to exist via small-scale or large-scale experimental methods. Examples: DIP, Biomolecular Interaction Network Database (BIND), Biological General Repository for Interaction Datasets (BioGRID), Human Protein Reference Database (HPRD), IntAct Molecular Interaction Database, Molecular Interactions Database (MINT), MIPS Protein Interaction Resource on Yeast (MIPS-MPact), and MIPS Mammalian Protein–Protein Interaction Database (MIPS-MPPI).

Meta-databases normally result from the integration of primary databases information, but can also collect some original data. Examples: Agile Protein Interactomes Dataserver (APID), The Microbial Protein Interaction Database (MPIDB), and Protein Interaction Network Analysis (PINA) platform, (GPS-Prot).

Prediction databases include many PPIs that are predicted using several techniques. Examples: Human Protein–Protein Interaction Prediction Database (PIPs), Interlogous Interaction Database (I2D), Known and Predicted Protein–Protein Interactions, and Unified Human Interactive (UniHI).

Interaction Networks

Information found in PPIs databases supports the construction of interaction networks. Although the PPI network of a given query protein can be represented in textbooks, diagrams of whole cell PPIs are frankly complex and difficult to generate.

One example of a manually produced molecular interaction map is the Kurt Kohn's 1999 map of cell cycle control. Drawing on Kohn's map, Schwikowski et al. in 2000 published a paper on PPIs in yeast, linking 1,548 interacting proteins determined by two-hybrid screening. They used a layered graph drawing method to find an initial placement of the nodes and then improved the layout using a force-based algorithm.

Bioinformatic tools have been developed to simplify the difficult task of visualizing molecular interaction networks and complement them with other types of data. For instance, Cytoscape is an open-source software widely used and lots of plugins are currently available. Pajek software is advantageous for the visualization and analysis of very large networks.

Identification of functional modules in PPI networks is an important challenge in bioinformatics. Functional modules means a set of proteins that are highly connected to each other in PPI network. It is almost similar problem as community detection in social networks. There are some methods such as Jactive modules and MoBaS. Jactive modules integrate PPI network and gene expression data where as MoBaS integrate PPI network and Genome Wide association Studies.

The awareness of the major roles of PPIs in numerous physiological and pathological processes has been driving the challenge of unravel many interactomes. Examples of published interactomes are the thyroid specific DREAM interactome and the PP1α interactome in human brain.

Protein-protein relationships are often the result of multiple types of interactions or are deduced from different approaches, including co-localization, direct interaction, suppressive genetic interaction, additive genetic interaction, physical association, and other associations.

Signed Interaction Networks

A

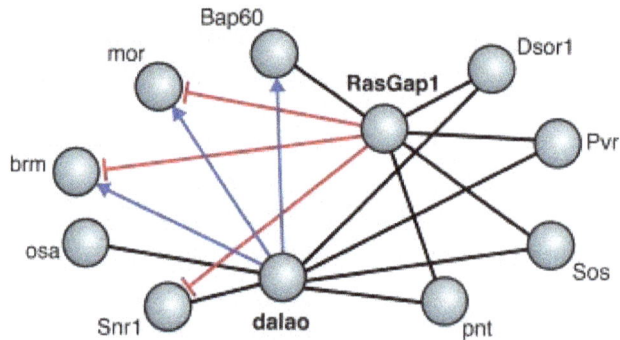

The protein protein interactions are displayed in a signed network
that describes what type of interactions that are taking place

Protein–protein interactions often result in one of the interacting proteins either being 'activated' or 'repressed'. Such effects can be indicated in a PPI network by "signs" (e.g. "activation" or "inhibition"). Although such attributes have been added to networks for a long time, Vinayagam et al. (2014) coined the term *Signed network* for them. Signed networks are often expressed by labeling the interaction as either positive or negative. A positive interaction is one where the interaction results in one of the proteins being activated. Conversely a negative interaction indicates that one of the proteins being inactivated.

Protein–protein interaction networks are often constructed as a result of lab experiments such as yeast two hybrid screens or 'affinity purification and subsequent mass spectrometry techniques. However these methods do not provide the layer of information needed in order to determine what type of interaction is present in order to be able to attribute signs to the network diagrams.

RNA Interference Screens

RNA interference (RNAi) screens (repression of individual proteins between transcription and translation) are one method that can be utilized in the process of providing signs to the protein-protein interactions. Individual proteins are repressed and the resulting phenotypes are analyzed. A correlating phenotypic relationship (i.e. where the inhibition of either of two proteins results in the same phenotype) indicates a positive, or activating relationship. Phenotypes that do not correlate (i.e. where the inhibition of either of two proteins results in two different phenotypes) indicate a negative or inactivating relationship. If protein A is dependent on protein B for activation then the inhibition of either protein A or B will result in a cell losing the service that is provided by protein A and the phenotypes will be the same for the inhibition of either A or B. If, however, protein A is inactivated by protein B then the phenotypes will differ depending on which protein is inhibited (inhibit protein B and it can no longer inactivate protein A leaving A active however inactivate A and there is nothing for B to activate since A is inactive and the phenotype changes). Multiple RNAi screens need to be performed in order to reliably appoint a sign to a given protein-protein interaction. Vinayagam et al. who devised this technique state that a minimum of nine RNAi screens are required with confidence increasing as one carries out more screens.

As Therapeutic Targets

Modulation of PPI is challenging and is receiving increasing attention by the scientific community. Several properties of PPI such as allosteric sites and hotspots, have been incorporated into drug-design strategies. The relevance of PPI as putative therapeutic targets for the development of new treatments is particularly evident in cancer, with several ongoing clinical trials within this area. The consensus among these promising targets is, nonetheless, denoted in the already available drugs on the market to treat a multitude of diseases. Examples are Titrobifan, inhibitor of the glycoprotein IIb/IIIa, used as a cardiovascular drug, and Maraviroc, inhibitor of the CCR5-gp120 interaction, used as anti-HIV drug. Recently, Amit Jaiswal and others were able to develop 30 peptides using protein–protein interaction studies to inhibit telomerase recruitment towards telomeres.

Biological Applications

The protein-protein interaction studies can be used for various biological applications such as:

- Identification of novel proteins
- Identification of function of unknown proteins
- Identification of new binding partners for known proteins
- Functional characterization of protein interactions
- Protein/antibody screening
- Biomarker discovery

Investigating Methods

There are many methods to investigate protein–protein interactions. Each of the approaches has its own strengths and weaknesses, especially with regard to the sensitivity and specificity of the method. A high sensitivity means that many of the interactions that occur in reality are detected by the screen. A high specificity indicates that most of the interactions detected by the screen are occurring in reality.

Biochemical Methods

- Co-immunoprecipitation is considered to be the gold standard assay for protein–protein interactions, especially when it is performed with endogenous (not overexpressed and not tagged) proteins. The protein of interest is isolated with a specific antibody. Interaction partners which stick to this protein are subsequently identified by Western blotting. Interactions detected by this approach are considered to be real. However, this method can only verify interactions between suspected interaction partners. Thus, it is not a screening approach. A note of caution also is that immunoprecipitation experiments reveal direct and indirect interactions. Thus, positive results may indicate that two proteins interact directly or may interact via one or more bridging molecules. This could include bridging proteins, nucleic acids (DNA or RNA), or other molecules.

- Bimolecular fluorescence complementation (BiFC) is a new technique in observing the interactions of proteins. Combining with other new techniques, this method can be used to screen protein–protein interactions and their modulators,DERB.

- Affinity electrophoresis as used for estimation of binding constants, as for instance in lectin affinity electrophoresis or characterization of molecules with specific features like glycan content or ligand binding.

- Pull-down assays are a common variation of immunoprecipitation and immunoelectrophoresis and are used identically, although this approach is more amenable to an initial screen for interacting proteins.

- Label transfer can be used for screening or confirmation of protein interactions and can provide information about the interface where the interaction takes place. Label transfer can also detect weak or transient interactions that are difficult to capture using other *in vitro* detection strategies. In a label transfer reaction, a known protein is tagged with a detectable label. The label is then passed to an interacting protein, which can then be identified by the presence of the label.

- The yeast two-hybrid screen investigates the interaction between artificial fusion proteins inside the nucleus of yeast. This approach can identify binding partners of a protein in an unbiased manner.

- Phage display, used for the high-throughput screening of protein interactions.

- *In-vivo* crosslinking of protein complexes using photo-reactive amino acid analogs was introduced in 2005 by researchers from the Max Planck Institute In this method, cells are grown with photoreactivediazirine analogs to leucine and methionine, which are incorporated into proteins. Upon exposure to ultraviolet light, the diazirines are activated and bind to interacting proteins that are within a few angstroms of the photo-reactive amino acid analog.

- Tandem affinity purification (TAP) method allows high throughput identification of protein interactions. In contrast to yeast two-hybrid approach the accuracy of the method can be compared to those of small-scale experiments and the interactions are detected within the correct cellular environment as by co-immunoprecipitation. However, the TAP tag method requires two successive steps of protein purification and consequently it can not readily detect transient protein–protein interactions. Recent genome-wide TAP experiments were performed by Krogan *et al.* and Gavin *et al.* providing updated protein interaction data for yeast organism.

- Chemical cross-linking is often used to "fix" protein interactions in place before trying to isolate/identify interacting proteins. Common crosslinkers for this application include the non-cleavable NHS-ester cross-linker, bissulfosuccinimidyl suberate (BS3); a cleavable version of BS3, dithiobis(sulfosuccinimidyl propionate) (DTSSP); and the imidoester cross-linker dimethyl dithiobispropionimidate (DTBP) that is popular for fixing interactions in ChIP assays.

- Chemical cross-linking followed by high mass MALDImass spectrometry can be used to analyze intact protein interactions in place before trying to isolate/identify interacting proteins. This method detects interactions among non-tagged proteins and is available from CovalX.

- SPINE (Strepprotein interaction experiment) uses a combination of reversible crosslinking with formaldehyde and an incorporation of an affinity tag to detect interaction partners *in vivo*.

- Quantitative immunoprecipitation combined with knock-down (QUICK) relies on co-immunoprecipitation, quantitative mass spectrometry (SILAC) and RNA interference (RNAi). This method detects interactions among endogenous non-tagged proteins.Thus, it has the same high confidence as co-immunoprecipitation. However, this method also depends on the availability of suitable antibodies.

- Proximity ligation assay (PLA) in situ is an immunohistochemical method utilizing so called PLA probes for detection of proteins, protein interactions and modifications. Each PLA probes comes with a unique short DNA strand attached to it and bind either to species specific primary antibodies or consist of directly DNA-labeled primary antibodies. When the PLA probes are in close proximity, the DNA strands can interact through a subsequent addition of two other circle-forming DNA oligonucleotides. After joining of the two added oligonucleotides by enzymatic ligation, they are amplified via rolling circle amplification using a polymerase. After the amplification reaction, several-hundredfold replication of the DNA circle has occurred and flurophore or enzyme labeled complementary oligonucleotide probes highlight the product. The resulting high concentration of fluorescence or cromogenic signal in each single-molecule amplification product is easily visible as a distinct bright spot when viewed with either in a fluorescence microscope or a standard brightfield microscope.

Biophysical and Theoretical Methods

- Bio-layer interferometry is a label-free technology for measuring biomolecular interactions (protein:protein or protein:small molecule). It is an optical analytical technique that analyzes the interference pattern of white light reflected from two surfaces: a layer of immobilized protein on the biosensor tip, and an internal reference layer. Any change in the number of molecules bound to the biosensor tip causes a shift in the interference pattern that can be measured in real-time, providing detailed information regarding the kinetics of association and dissociation of the two molecule molecules as well as the affinity constant for the protein interaction (k_a, k_d and K_d). Due to sensor configuration, the technique is highly amenable to both purified and crude samples as well as high throughput screening experiments. The detection method can also be used to determine the molar concentration of analytes. Bio-layer interferometry was pioneered by the founders of ForteBio, an instrument manufacturer with headquarters in Menlo Park California.

- Dual polarisation interferometry (DPI) can be used to measure protein–protein interactions. DPI provides real-time, high-resolution measurements of molecular size, density and mass. While tagging is not necessary, one of the protein species must be immobilized on the surface of a waveguide. As well as kinetics and affinity, conformational changes during interaction can also be quantified.

- Static light scattering (SLS) measures changes in the Rayleigh scattering of protein complexes in solution and can characterize both weak and strong interactions without labeling or immobilization of the proteins or other biomacromolecule. The composition-gradient, multi-angle static light scattering (CG-MALS) measurement mixes a series of aliquots of different concentrations or compositions, measures the effect of the changes in light scattering as a result of the interaction, and fits the correlated light scattering changes with concentration to a series of association models in order to find the best-fit descriptor. Weak, non-specific interactions are typically characterized via the second virial coefficient. For specific binding, this type of analysis can determine the stoichiometry and equilibrium association constant(s) of one or more associated complexes, including challenging systems such as those that exhibit simultaneous homo- and hetero-association, multi-valent interactions and cooperativity.

- Dynamic light scattering (DLS), also known as quasielastic light scattering (QELS), or photon correlation spectroscopy, processes the time-dependent fluctuations in scattered light intensity to yield the hydrodynamic radius of particles in solution. The hydrodynamic radius is the radius of a solid sphere with the same translational diffusion coefficient as that measured for the sample particle. As proteins associate, the average hydrodynamic radius of the solution increases. Application of the Method of Continuous Variation, otherwise known as the Job plot, with the solution hydrodynamic radius as the observable, enables *in vitro* determination of K_d, complex stoichiometry, complex hydrodynamic radius, and the $\Delta H°$ and $\Delta S°$ of protein–protein interactions. This technique does not entail immobilization or labeling. Transient and weak interactions can be characterized. Relative to static light scattering, which is based upon the absolute intensity of scattered light, DLS is insensitive to background light from the walls of containing structures. This insensitivity permits DLS measurements from 1 μL volumes in 1536 well plates, and lowers sample requirements into the femtomole range. This technique is also suitable for screening of buffer components and/or small molecule inhibitors/effectors.

- Surface plasmon resonance is the most common label-free technique for the measurement of biomolecular interactions. SPR instruments measure the change in the refractive index of light reflected from a metal surface (the "biosensor"). Binding of biomolecules to the other side of this surface leads to a change in the refractive index which is proportional to the mass added to the sensor surface. In a typical application, one binding partner (the "ligand", often a protein) is immobilized on the biosensor and a solution with potential binding partners (the "analyte") is channelled over this surface. The build-up of analyte over time allows to quantify on rates (kon), off rates (koff), dissociation constants (Kd) and, in some applications, active concentrations of the analyte. Several different vendors offer SPR-based devices. Best known are Biacore instruments which were the first commercially available.

- Fluorescence polarization/anisotropy can be used to measure protein–protein or protein–ligand interactions. Typically one binding partner is labeled with a fluorescence probe (although sometimes intrinsic protein fluorescence from tryptophan can be used) and the sample is excited with polarized light. The increase in the polarization of the fluorescence upon binding of the labeled protein to its binding partner can be used to calculate the binding affinity.

- With fluorescence correlation spectroscopy, one protein is labeled with a fluorescent dye and the other is left unlabeled. The two proteins are then mixed and the data outputs the fraction of the labeled protein that is unbound and bound to the other protein, allowing you to get a measure of K_D and binding affinity. You can also take time-course measurements to characterize binding kinetics. FCS also tells you the size of the formed complexes so you can measure the stoichiometry of binding. A more powerful methods is fluorescence cross-correlation spectroscopy (FCCS) that employs double labeling techniques and cross-correlation resulting in vastly improved signal-to-noise ratios over FCS. Furthermore, the two-photon and three-photon excitation practically eliminates photobleaching effects and provide ultra-fast recording of FCCS or FCS data.

- Fluorescence resonance energy transfer (FRET) is a common technique when observing the interactions of only two different proteins.

- Protein activity determination by NMR multi-nuclear relaxation measurements, or 2D-FT NMR spectroscopy in solutions, combined with nonlinear regression analysis of NMR relaxation or 2D-FT spectroscopy data sets. Whereas the concept of water activity is widely known and utilized in the applied biosciences, its complement—the protein activity which quantitates protein–protein interactions—is much less familiar to bioscientists as it is more difficult to determine in dilute solutions of proteins; protein activity is also much harder to determine for concentrated protein solutions when protein aggregation, not merely transient protein association, is often the dominant process.

- Protein–protein docking, the prediction of protein–protein interactions based only on the three-dimensional protein structures from X-ray diffraction of protein crystals might not be satisfactory.

- Isothermal titration calorimetry (ITC), is considered as the most quantitative technique available for measuring the thermodynamic properties of protein–protein interactions and is becoming a necessary tool for protein–protein complex structural studies. This technique relies upon the accurate measurement of heat changes that follow the interaction of protein molecules in solution, without the need to label or immobilize the binding partners, since the absorption or production of heat is an intrinsic property of virtually all biochemical reactions. ITC provides information regarding the stoichiometry, enthalpy, entropy, and binding kinetics between two interacting proteins.

- Microscale thermophoresis (MST), is a new method that enables the quantitative analysis of molecular interactions in solution at the microliter scale. The technique is based on the thermophoresis of molecules, which provides information about molecule size, charge and hydration shell. Since at least one of these parameters is typically affected upon binding, the method can be used for the analysis of each kind of biomolecular interaction or modification. The method works equally well in standard buffers and biological liquids like blood or cell-lysate. It is a free solution method which does not need to immobilize the binding partners. MST provides information regarding the binding affinity, stoichiometry, competition and enthalpy of two or more interacting proteins.

References

- Mika S, Rost B (2006). "Protein–Protein Interactions More Conserved within Species than across Species". PLoS Computational Biology. 2 (7): e79. PMC 1513270. PMID 16854211. doi:10.1371/journal.pcbi.0020079

- Bruce A, Johnson A, Lewis J, Raff M, Roberts K, Walter P (2002). Molecular biology of the cell (4th ed.). New York: Garland Science. ISBN 0-8153-3218-1

- Costanzo M, Baryshnikova A, Bellay J, et al. (2010-01-22). "The genetic landscape of a cell". Science. 327 (5964): 425–431. PMID 20093466. doi:10.1126/science.1180823

- Mashaghi, A.; et al. (2004). "Investigation of a protein complex network". European Physical Journal. 41 (1): 113–121. doi:10.1140/epjb/e2004-00301-0

- Bellucci M (2010). Protein-Protein Interactions: A tool kit to puzzle out functional networks. VDM Verlag Dr. Müller. ISBN 978-3-639-31160-0

- Bruggeman, F J; H V Westerhoff (2006). "The nature of systems biology". TRENDS in Microbiology. 15 (1): 45–50. PMID 17113776. doi:10.1016/j.tim.2006.11.003

- Romanuk, T.; et al. (2010). "Maintenance of positive diversity-stability relations along a gradient of environmental stress". PLoS ONE. 5: e10378. PMC 2860506. PMID 20436913. doi:10.1371/journal.pone.0010378

- Kohl M, Wiese S, Warscheid B (2011). "Cytoscape: Software for Visualization and Analysis of Biological Networks". Data Mining in Proteomics. Methods in Molecular Biology. 696. pp. 291–303. ISBN 978-1-60761-986-4. doi:10.1007/978-1-60761-987-1_18

- Rajagopala SV.; et al. (2011). "The protein interaction map of bacteriophage lambda". BMC Microbiol. 11: 213. PMC 3224144. PMID 21943085. doi:10.1186/1471-2180-11-213

- Berridge, M.J. (2012). "Cell Signalling Biology: Module 6 – Spatial and Temporal Aspects of Signalling". Biochemical Journal. doi:10.1042/csb0001006

- Baianu IC, Pressen H, Kumosinski TF (1993). "NMR Principles and Applications to Protein Structure, Activity and Hydration". Advanced Techniques, Structures and Applications. Physical Chemistry of Food Processes, Volume II. New York: Van Nostrand-Reinhold. pp. 338–420. ISBN 0-442-00582-2

- Hagen, N; Bayer, K; Roesch, K; Schindler, M (2014). "The intra viral protein interaction network of hepatitis C virus". Molecular & Cellular Proteomics. 13 (7): 1676–89. PMID 24797426. doi:10.1074/mcp.M113.036301

- Badal VD, Kundrotas PJ, Vakser IA (December 2015). "Text Mining for Protein Docking". PLoS Computational Biology. 11 (12): e1004630. PMC 4674139. PMID 26650466. doi:10.1371/journal.pcbi.1004630

Detection Methods of Protein-Protein Interaction

Two-hybrid screening is a technique which studies the protein-protein interaction and the protein and DNA interaction. Affinity chromatography, immunoprecipitation, tandem affinity purification and protein mass spectrometry are some of the topics focused upon in this section. The chapter strategically encompasses and incorporates the major components and key concepts of protein-protein interaction, providing a complete understanding.

Two Hybrid System

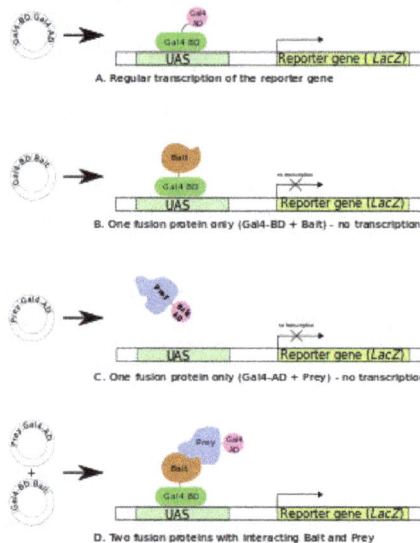

A. Regular transcription of the reporter gene

B. One fusion protein only (Gal4-BD + Bait) - no transcription

C. One fusion protein only (Gal4-AD + Prey) - no transcription

D. Two fusion proteins with interacting Bait and Prey

Overview of two-hybrid assay, checking for interactions between two proteins, called here *Bait* and *Prey*.

In the above image A, B, C and D means:

A. The*Gal4* transcription factor gene produces a two-domain protein (*BD* and *AD*) essential for transcription of the reporter gene (*LacZ*).

B,C. Two fusion proteins are prepared: *Gal4BD+Bait* and *Gal4AD+Prey*. Neither of them are usually sufficient to initiate transcription (of the reporter gene) alone.

D. When both fusion proteins are produced and the Bait part of the first fusion protein interacts with the Prey part of the second, transcription of the reporter gene occurs.

Two-hybrid screening (originally known as yeast two-hybrid system or Y2H) is a molecular biology technique used to discover protein–protein interactions (PPIs) and protein–DNA interactions by testing for physical interactions (such as binding) between two proteins or a single protein and a DNA molecule, respectively.

The premise behind the test is the activation of downstreamreporter gene(s) by the binding of a transcription factor onto an upstream activating sequence (UAS). For two-hybrid screening, the transcription factor is split into two separate fragments, called the DNA-binding domain (DBD or often also abbreviated as BD) and activating domain (AD). The BD is the domain responsible for binding to the UAS and the AD is the domain responsible for the activation of transcription. The Y2H is thus a protein-fragment complementation assay.

History

Pioneered by Stanley Fields and Ok-Kyu Song in 1989, the technique was originally designed to detect protein–protein interactions using the GAL4 transcriptional activator of the yeast *Saccharomyces cerevisiae*. The GAL4 protein activated transcription of a protein involved in galactose utilization, which formed the basis of selection. Since then, the same principle has been adapted to describe many alternative methods, including some that detect protein–DNA interactions or DNA-DNA interactions, as well as methods that use different host organisms such as *Escherichia coli* or mammalian cells instead of yeast.

Basic Premise

The key to the two-hybrid screen is that in most eukaryotic transcription factors, the activating and binding domains are modular and can function in proximity to each other without direct binding. This means that even though the transcription factor is split into two fragments, it can still activate transcription when the two fragments are indirectly connected.

The most common screening approach is the yeast two-hybrid assay. In this approach the researcher knows where each prey is located on the used medium (agar plates). Millions of potential interactions in several organisms have been screened in the latest decade using high-throughput screening systems (often using robots) and over thousands of interactions have been detected and categorized in databases as BioGRID. This system often utilizes a genetically engineered strain of yeast in which the biosynthesis of certain nutrients (usually amino acids or nucleic acids) is lacking. When grown on media that lacks these nutrients, the yeast fail to survive. This mutant yeast strain can be made to incorporate foreign DNA in the form of plasmids. In yeast two-hybrid screening, separate bait and prey plasmids are simultaneously introduced into the mutant yeast strain or a mating strategy is used to get both plasmids in one host cell.

The second high-throughput approach is the library screening approach. In this set up the bait and prey harboring cells are mated in a random order. After mating and selecting surviving cells on selective medium the scientist will sequence the isolated plasmids to see which prey (DNA sequence) is interacting with the used bait. This approach has a lower rate of reproducibility and tends to yield higher amounts of false positives compared to the matrix approach.

Plasmids are engineered to produce a protein product in which the DNA-binding domain (BD) fragment is fused onto a protein while another plasmid is engineered to produce a protein product in which the activation domain (AD) fragment is fused onto another protein. The protein fused to the BD may be referred to as the bait protein, and is typically a known protein the investigator is using to identify new binding partners. The protein fused to the AD may be referred to as the prey protein and can be either a single known protein or a library of known or unknown proteins. In this context, a library may consist of a collection of protein-encoding sequences that represent all the proteins expressed in a particular organism or tissue, or may be generated by synthesising random DNA sequences. Regardless of the source, they are subsequently incorporated into the protein-encoding sequence of a plasmid, which is then transfected into the cells chosen for the screening method. This technique, when using a library, assumes that each cell is transfected with no more than a single plasmid and that, therefore, each cell ultimately expresses no more than a single member from the protein library.

If the bait and prey proteins interact (i.e., bind), then the AD and BD of the transcription factor are indirectly connected, bringing the AD in proximity to the transcription start site and transcription of reporter gene(s) can occur. If the two proteins do not interact, there is no transcription of the reporter gene. In this way, a successful interaction between the fused protein is linked to a change in the cell phenotype.

Fixed Domains

In any study, some of the protein domains, those under investigation, will be varied according to the goals of the study whereas other domains, those that are not themselves being investigated, will be kept constant. For example, in a two-hybrid study to select DNA-binding domains, the DNA-binding domain, BD, will be varied while the two interacting proteins, the bait and prey, must be kept constant to maintain a strong binding between the BD and AD. There are a number of domains from which to choose the BD, bait and prey and AD, if these are to remain constant. In protein–protein interaction investigations, the BD may be chosen from any of many strong DNA-binding domains such as Zif268. A frequent choice of bait and prey domains are residues 263–352 of yeast Gal11P with a N342V mutation and residues 58–97 of yeast Gal4, respectively. These domains can be used in both yeast- and bacterial-based selection techniques and are known to bind together strongly.

The AD chosen must be able to activate transcription of the reporter gene, using the cell's own transcription machinery. Thus, the variety of ADs available for use in yeast-based techniques may not be suited to use in their bacterial-based analogues. The herpes simplex virus-derived AD, VP16 and yeast Gal4 AD have been used with success in yeast whilst a portion of the α-subunit of *E. coli* RNA polymerase has been utilised in *E. coli*-based methods.

Whilst powerfully activating domains may allow greater sensitivity towards weaker interactions, conversely, a weaker AD may provide greater stringency.

Construction of Expression Plasmids

A number of engineered genetic sequences must be incorporated into the host cell to perform two-hybrid analysis or one of its derivative techniques. The considerations and methods used in the construction and delivery of these sequences differ according to the needs of the assay and the organism chosen as the experimental background.

There are two broad categories of hybrid library: random libraries and cDNA-based libraries. A cDNA library is constituted by the cDNA produced through reverse transcription of mRNA collected from specific cells of types of cell. This library can be ligated into a construct so that it is attached to the BD or AD being used in the assay. A random library uses lengths of DNA of random sequence in place of these cDNA sections. A number of methods exist for the production of these random sequences, including cassette mutagenesis. Regardless of the source of the DNA library, it is ligated into the appropriate place in the relevant plasmid/phagemid using the appropriate restriction endonucleases.

E. Coli-specific Considerations

By placing the hybrid proteins under the control of IPTG-inducible *lac* promoters, they are expressed only on media supplemented with IPTG. Further, by including different antibiotic resistance genes in each genetic construct, the growth of non-transformed cells is easily prevented through culture on media containing the corresponding antibiotics. This is particularly important for counter selection methods in which a *lack* of interaction is needed for cell survival.

The reporter gene may be inserted into the *E. coli* genome by first inserting it into an episome, a type of plasmid with the ability to incorporate itself into the bacterial cell genome with a copy number of approximately one per cell.

The hybrid expression phagemids can be electroporated into *E. coli* XL-1 Blue cells which after amplification and infection with VCS-M13 helper phage, will yield a stock of library phage. These phage will each contain one single-stranded member of the phagemid library.

Recovery of Protein Information

Once the selection has been performed, the primary structure of the proteins which display the appropriate characteristics must be determined. This is achieved by retrieval of the protein-encoding sequences (as originally inserted) from the cells showing the appropriate phenotype.

E. Coli

The phagemid used to transform *E. coli* cells may be "rescued" from the selected cells by infecting them with VCS-M13 helper phage. The resulting phage particles that are produced contain the single-stranded phagemids and are used to infect XL-1 Blue cells. The double-stranded phagemids are subsequently collected from these XL-1 Blue cells, essentially reversing the process used to produce the original library phage. Finally, the DNA sequences are determined through dideoxy sequencing.

Controlling Sensitivity

The *Escherichia coli*-derived Tet-R repressor can be used in line with a conventional reporter gene and can be controlled by tetracycline or doxicycline (Tet-R inhibitors). Thus the expression of Tet-R is controlled by the standard two-hybrid system but the Tet-R in turn controls (represses) the expression of a previously mentioned reporter such as *HIS3*, through its Tet-R promoter. Tetracycline or its derivatives can then be used to regulate the sensitivity of a system utilising Tet-R.

Sensitivity may also be controlled by varying the dependency of the cells on their reporter genes. For example, this may be affected by altering the concentration of histidine in the growth medium for *his3*-dependent cells and altering the concentration of streptomycin for *aadA* dependent cells. Selection-gene-dependency may also be controlled by applying an inhibitor of the selection gene at a suitable concentration. 3-Amino-1,2,4-triazole (3-AT) for example, is a competitive inhibitor of the *HIS3*-gene product and may be used to titrate the minimum level of *HIS3* expression required for growth on histidine-deficient media.

Sensitivity may also be modulated by varying the number of operator sequences in the reporter DNA.

Non-fusion Proteins

A third, non-fusion protein may be co-expressed with two fusion proteins. Depending on the investigation, the third protein may modify one of the fusion proteins or mediate or interfere with their interaction.

Co-expression of the third protein may be necessary for modification or activation of one or both of the fusion proteins. For example, *S. cerevisiae* possesses no endogenous tyrosine kinase. If an investigation involves a protein that requires tyrosine phosphor-ylation, the kinase must be supplied in the form of a tyrosine kinase gene.

The non-fusion protein may mediate the interaction by binding both fusion proteins simultaneously, as in the case of ligand-dependent receptor dimerization.

For a protein with an interacting partner, its functional homology to other proteins may be assessed by supplying the third protein in non-fusion form, which then may or may not compete with the fusion-protein for its binding partner. Binding between the third protein and the other fusion protein will interrupt the formation of the reporter expression activation complex and thus reduce reporter expression, leading to the distinguishing change in phenotype.

Split-ubiquitin Yeast Two-hybrid

One limitation of classic yeast two-hybrid screens is that they are limited to soluble proteins. It is therefore impossible to use them to study the protein–protein interactions between insoluble integral membrane proteins. The split-ubiquitin system provides a method for overcoming this limitation. In the split-ubiquitin system, two integral membrane proteins to be studied are fused to two different ubiquitin moieties: a C-terminal ubiquitin moiety ("Cub", residues 35–76) and an N-terminal ubiquitin moiety ("Nub", residues 1–34). These fused proteins are called the bait and prey, respectively. In addition to being fused to an integral membrane protein, the Cub moiety is also fused to a transcription factor (TF) that can be cleaved off by ubiquitin specific proteases. Upon bait–prey interaction, Nub and Cub-moieties assemble, reconstituting the split-ubiquitin. The reconstituted split-ubiquitin molecule is recognized by ubiquitin specific proteases, which cleave off the transcription factor, allowing it to induce the transcription of reporter genes.

Fluorescent Two-hybrid Assay

Zolghadr and co-workers presented a fluorescent two-hybrid system that uses two hybrid proteins that are fused to different fluorescent proteins as well as LacI, the lac repressor. The structure of the fusion proteins looks like this: FP2-LacI-bait and FP1-prey where the bait and prey proteins interact and bring the fluorescent proteins (FP1 = GFP, FP2=m-Cherry) in close proximity at the binding site of the LacI protein in the host cell genome. The system can also be used to screen for inhibitors of protein–protein interactions.

Enzymatic Two-hybrid Systems: KISS

While the original Y2H system used a reconstituted transcription factor, other systems create enzymatic activities to detect PPIs. For instance, the KInase Substrate Sensor ("KISS"), is a mammalian two-hybrid approach has been designed to map intracellular PPIs. Here, a bait protein is fused to a kinase-containing portion of TYK2 and a prey is coupled to a gp130cytokine receptor fragment. When bait and prey interact, TYK2 phosphorylates STAT3 docking sites on the prey chimera, which ultimately leads to activation of a reporter gene.

One-, Three- and One-two-hybrid Variants

One-hybrid

The one-hybrid variation of this technique is designed to investigate protein–DNA interactions and uses a single fusion protein in which the AD is linked directly to the binding domain. The binding domain in this case however is not necessarily of fixed sequence as in two-hybrid protein–protein analysis but may be constituted by a library. This library can be selected against the desired target sequence, which is inserted in the promoter region of the reporter gene construct. In a positive-selection system, a binding domain that successfully binds the UAS and allows transcription is thus selected.

Note that selection of DNA-binding domains is not necessarily performed using a one-hybrid system, but may also be performed using a two-hybrid system in which the binding domain is varied and the bait and prey proteins are kept constant.

Three-hybrid

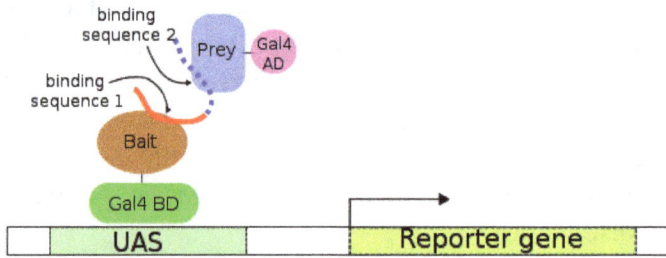

Overview of three-hybrid assay.

RNA-protein interactions have been investigated through a three-hybrid variation of the two-hybrid technique. In this case, a hybrid RNA molecule serves to adjoin together the two protein fusion domains—which are not intended to interact with each other but rather the intermediary RNA molecule (through their RNA-binding domains).

One-two-hybrid

Simultaneous use of the one- and two-hybrid methods (that is, simultaneous protein–protein and protein–DNA interaction) is known as a one-two-hybrid approach and expected to increase the stringency of the screen.

Host Organism

Although theoretically, any living cell might be used as the background to a two-hybrid analysis, there are practical considerations that dictate which is chosen. The chosen cell line should be relatively cheap and easy to culture and sufficiently robust to withstand application of the investigative methods and reagents. The latter is especially important for doing high-throughput studies. Therefore the yeast *S. cerevisiae* has been the main host organism for two-hybrid studies. However it is not always the ideal system to study interacting proteins from other organisms. Yeast cells often do not have the same post translational modifications, have a different codon use or lack certain proteins that are important for the correct expression of the proteins. To cope with these problems several novel two-hybrid systems have been developed. Depending on the system used agar plates or specific growth medium is used to grow the cells and allow selection for interaction. The most common used method is the agar plating one where cells are plated on selective medium to see of interaction takes place. Cells that have no interaction proteins should not survive on this selective medium.

S. Cerevisiae

The yeast *S. cerevisiae* was the model organism used during the two-hybrid technique's inception. It is commonly known as the Y2H system. It has several characteristics that make it a robust organism to host the interaction, including the ability to form tertiary protein structures, neutral internal pH, enhanced ability to form disulfide bonds and reduced-state glutathione among other cytosolic buffer factors, to maintain a hospitable internal environment. The yeast model can be manipulated through non-molecular techniques and its complete genome sequence is known. Yeast systems are tolerant of diverse culture conditions and harsh chemicals that could not be applied to mammalian tissue cultures.

A number of yeast strains have been created specifically for Y2H screens, e.g. Y187 and AH109, both produced by Clontech. Yeast strains R2HMet and BK100 have also been used.

Candida Albicans

C. albicans is a yeast with a particular feature: it translates the CUG codon into serine rather than leucine. Due to this different codon usage it is difficult to use the model system *S. cerevisiae* as a Y2H to check for protein-protein interactions using *C. albicans* genes. To provide a more native environment a *C. albicans* two-hybrid (C2H) system was developed. With this system protein-protein interactions can be studied in *C. albicans* itself.

E. Coli

Bacterial *E. coli*-based two hybrid methods (abbreviated as B2H) have several characteristics that may make them preferable to yeast-based homologues. The higher transformation efficiency and faster rate of growth lends *E. coli* to the use of larger libraries (in excess of 10^8). A low false positive rate of approximately $3x10^{-8}$, the absence of requirement for a nuclear localisation signal to be included in the protein sequence and the ability to study proteins that would be toxic to yeast may also be major factors to consider when choosing an experimental background organism.

It may be of note that the methylation activity of certain *E. coli*DNA methyltransferase proteins may interfere with some DNA-binding protein selections. If this is anticipated, the use of an *E. coli* strain that is defective for a particular methyltransferase may be an obvious solution. Important to mention is that bacteria are prokaryotic organisms and when studying eukaryotic protein-protein interactions (e.g. human proteins) the results need to be carefully approached.

Mammalian Cells

In recent years a mammalian two hybrid (M2H) system has been designed to study mammalian protein-protein interactions in a cellular environment that closely mimics the native protein environment Transiently transfected mammalian cells are used in this system to find protein-protein interactions. Using a mammalian cell line to study mammalian protein-protein interactions gives the advantage of working in a more native context. The post-translational modifications, phosphorylation, acylation and glycosylation are similar. The intracellular localization of the proteins is also more correct compared to using a yeast two hybrid system.It is also possible with the mammalian two-hybrid system to study signal inputs. Another big advantage is that results can be obtained wishing 48 hours after transfection.

Arabidopsis Thaliana

In 2005 a two hybrid system in plants was developed. Using protoplasts of *A. thaliana* protein-protein interactions can be studied in plants. This way the interactions can be studied in their native context. In this system the GAL4 AD and BD are under the control of the strong 35S promoter. Interaction is measured using a GUS reporter. In order to enable a high-throughput screening the vectors were made gateway compatible. The system is known as the protoplast two hybrid (P2H) system.

Aplysia Californica

The sea hare *A californica* is a model organism in neurobiology to study among others the molecular mechanisms of long-term memory. To study interactions, important in neurology, in a more native environment a two-hybrid system has been developed in *A californica* neurons. A GAL4 AD and BD are used in this system.

Bombyx Mori

An insect two-hybrid (I2H) system was developed in a silkworm cell line from the larva or caterpillar of the domesticated silk moth, *Bombyx mori* (BmN4 cells). This system uses the GAL4 BD and the activation domain of mouse NF-κB P65. Both are under the control of the OpIE2 promoter.

Applications

Determination of Sequences Crucial for Interaction

By changing specific amino acids by mutating the corresponding DNA base-pairs in the plasmids used, the importance of those amino acid residues in maintaining the interaction can be determined.

After using bacterial cell-based method to select DNA-binding proteins, it is necessary to check the specificity of these domains as there is a limit to the extent to which the bacterial cell genome can act as a sink for domains with an affinity for other sequences (or indeed, a general affinity for DNA).

Drug and Poison Discovery

Protein–protein signalling interactions pose suitable therapeutic targets due to their specificity and pervasiveness. The random drug discovery approach uses compound banks that comprise random chemical structures, and requires a high-throughput method to test these structures in their intended target.

The cell chosen for the investigation can be specifically engineered to mirror the molecular aspect that the investigator intends to study and then used to identify new human or animal therapeutics or anti-pest agents.

Determination of Protein Function

By determination of the interaction partners of unknown proteins, the possible functions of these new proteins may be inferred. This can be done using a single known protein against a library of unknown proteins or conversely, by selecting from a library of known proteins using a single protein of unknown function.

Zinc Finger Protein Selection

To select zinc finger proteins (ZFPs) for protein engineering, methods adapted from the two-hybrid screening technique have been used with success. A ZFP is itself a DNA-binding protein used in the construction of custom DNA-binding domains that bind to a desired DNA sequence.

By using a selection gene with the desired target sequence included in the UAS, and randomising the relevant amino acid sequences to produce a ZFP library, cells that host a DNA-ZFP interaction with the required characteristics can be selected. Each ZFP typically recognises only 3–4 base pairs, so to prevent recognition of sites outside the UAS, the randomised ZFP is engineered into a 'scaffold' consisting of another two ZFPs of constant sequence. The UAS is thus designed to include the target sequence of the constant scaffold in addition to the sequence for which a ZFP is selected.

A number of other DNA-binding domains may also be investigated using this system.

Strengths

- Two-hybrid screens are low-tech; they can be carried out in any lab without sophisticated equipment.

- Two-hybrid screens can provide an important first hint for the identification of interaction partners.

- The assay is scalable, which makes it possible to screen for interactions among many proteins. Furthermore, it can be automated, and by using robots many proteins can be screened against thousands of potentially interacting proteins in a relatively short time. Two types of large screens are used: the library approach and the matrix approach.

- Yeast two-hybrid data can be of similar quality to data generated by the alternative approach of coaffinity purification followed by mass spectrometry (AP/MS).

Weaknesses

- The main criticism applied to the yeast two-hybrid screen of protein–protein interactions are the possibility of a high number of false positive (and false negative) identifications. The exact rate of false positive results is not known, but earlier estimates were as high as 70%. This also, partly, explains the often found very small overlap in results when using a (high throughput) two-hybrid screening, especially when using different experimental systems.

The reason for this high error rate lies in the characteristics of the screen:

- Certain assay variants overexpress the fusion proteins which may cause unnatural protein concentrations that lead to unspecific (false) positives.

- The hybrid proteins are fusion proteins; that is, the fused parts may inhibit certain interactions, especially if an interaction takes place at the N-terminus of a test protein (where the DNA-binding or activation domain is typically attached).

- An interaction may not happen in yeast, the typical host organism for Y2H. For instance, if a bacterial protein is tested in yeast, it may lack a chaperone for proper folding that is only present in its bacterial host. Moreover, a mammalian protein is sometimes not correctly modified in yeast (e.g., missing phosphorylation), which can also lead to false results.

- The Y2H takes place in the nucleus. If test proteins are not localized to the nucleus (because they have other localization signals) two interacting proteins may be found to be non-interacting.

- Some proteins might specifically interact when they are co-expressed in the yeast, although in reality they are never present in the same cell at the same time. However, in most cases it cannot be ruled out that such proteins are indeed expressed in certain cells or under certain circumstances.

Each of these points alone can give rise to false results. Due to the combined effects of all error sources yeast two-hybrid have to be interpreted with caution. The probability of generating false positives means that all interactions should be confirmed by a high confidence assay, for example co-immunoprecipitation of the endogenous proteins, which is difficult for large scale protein–protein interaction data. Alternatively, Y2H data can be verified using multiple Y2H variants or bioinformatics techniques. The latter test whether interacting proteins are expressed at the same time, share some common features (such as gene ontology annotations or certain network topologies), have homologous interactions in other species.

The Two hybrid system uses this approach for detecting interaction among different proteins. It has been found that there is no need of covalent attachment between the two domains for the transcriptional activity. Transcription can be activated if the two domains can be brought together, which is done with the help of two interacting proteins attached with each domain. This aspect is also used in order to check whether two different proteins interact among themselves or not, if they interact then there will be the expression of reporter gene otherwise no expression.

The Construction of Two hybrid system includes creation of two different gene containing vectors i.e

1. Gene of DNA binding domain fused Bait protein (X- Bait protein)

2. Gene of Transcriptional activation domain fused Prey protein (Y-Prey protein)

These hybrid constructs are expressed in Y east cells containing Reporter genes. If the Protein X (Bait protein) and Protein Y (Prey protein) interact, they will bring the transcription a ctivation domain into close vicinity with the DNA binding domain, thus triggering the transcriptional process by allowing the RNA polymerase to bind and initiate transcription which is detected by the expression of the specific r eporter gene (e.g. color change in case of gal1-lacZ - the beta- galactosidase gene).

Steps for detecting Protein-protein interaction.

1. A bait vector is created consisting of selected protein to act as a bait (protein of interest against which interaction is to be checked) and DNA binding domain.

The product of this vector is a Fusion protein consisting of bait protein and DNA binding domain.

2. Another Prey vector is created consisting of transcription activation domain and a specific protein whose interaction with respect to bait protein is to be checked. The product of this vector is also a fusion product of both prey protein and transcription activation domain.

3. The Fusion protein of bait vector binds to the specific promoter region of the reporter gene with the help of DNA binding domain but unable to initiate transcription as lacks activation domain.

4. The prey protein fused with transcription activation domain, if interacts compatibly with bait protein then the activation factor is able to locate itself in the vicinity of reporter gene and able to initiate the transcription by allowing RNA polymerase to act.

5. Thus, the change in color i.e. expression of the reporter gene clears that the bait protein is in interaction with prey protein and thus conclusion can be made about their functional activity further.

The Two hybrid system can be used to check the interaction of a single bait protein against thousands of prey proteins as only those cells would grow into which there is proper interaction taking place and allowing the selected gene essential for growth to be expressed.

Affinity Chromatography

Affinity chromatography is a method of separating biochemical mixtures based on a highly specific interaction between antigen and antibody, enzyme and substrate, or receptor and ligand. It is a type of chromatographiclaboratory technique used for purifying biological molecules within a mixture by exploiting molecular properties. Biological macromolecules such as enzymes and other proteins, interact with other molecules with high specificity through several different types of bonds and interaction. Such interactions including hydrogen bonding, ionic interaction, disulfide bridges, hydrophobic interaction, and more. The high selectivity of affinity chromatography is caused by allowing the desired molecule to interact with the stationary phase and be bound within the column in order to be separated from the undesired material which will not interact and elute first. The molecules no longer needed are first washed away with a buffer while the desired proteins are let go in the presence of the eluting solvent (of higher salt concentration). This process creates a competitive interaction between the desired protein and the immobilized stationary molecules, which eventually lets the now highly purified proteins be released.

Uses

Affinity chromatography can be used to:

- Purify and concentrate a substance from a mixture into a buffering solution

- Reduce the amount of a the unwanted substances in a mixture

- Discern what biological compounds bind to a particular substance

- Purify and concentrate an enzyme solution.

Principle

The stationary phase is typically a gel matrix, often of agarose; a linear sugar molecule derived from algae. Usually the starting point is an undefined heterogeneous group of molecules in solution, such as a cell lysate, growth medium or blood serum. The molecule of interest will have a well known and defined property, and can be exploited during the affinity purification process. The process itself can be thought of as an entrapment, with the target molecule becoming trapped on a solid or stationary phase or medium. The other molecules in the mobile phase will not become trapped as they do not possess this property. The stationary phase can then be removed from the mixture, washed and the target molecule released from the entrapment in a process known as dialysis. Possibly the most common use of affinity chromatography is for the purification of recombinant proteins.

Batch and Column Setups

Add mixture to column.
Discard flow through.

Add wash to column.
Discard flow through.

Add elution buffer to column.
Retain Flow through.

Column chromatography

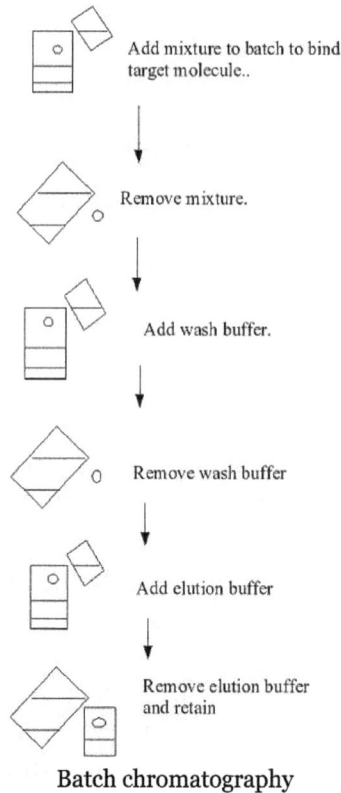

Add mixture to batch to bind
target molecule..

Remove mixture.

Add wash buffer.

Remove wash buffer

Add elution buffer

Remove elution buffer
and retain

Batch chromatography

Binding to the solid phase may be achieved by column chromatography whereby the solid medium is packed onto a column, the initial mixture run through the column to allow setting, a wash buffer run through the column and the elution buffer subsequently applied to the column and collected. These steps are usually done at ambient pressure. Alternatively, binding may be achieved using a batch treatment, for example, by adding the initial mixture to the solid phase in a vessel, mixing, separating the solid phase, removing the liquid phase, washing, re-centrifuging, adding the elution buffer, re-centrifuging and removing the elute.

Sometimes a hybrid method is employed such that the binding is done by the batch method, but the solid phase with the target molecule bound is packed onto a column and washing and elution are done on the column.

The ligands used in affinity chromatography are obtained from both organic and inorganic sources. Examples of biological sources are serum proteins, lectins and antibodies. Inorganic sources as moronic acts, metal chelates and triazine dyes.

A third method, expanded bed absorption, which combines the advantages of the two methods mentioned above, has also been developed. The solid phase particles are placed in a column where liquid phase is pumped in from the bottom and exits at the top. The gravity of the particles ensure that the solid phase does not exit the column with the liquid phase.

Affinity columns can be eluted by changing salt concentrations, pH, pI, charge and ionic strength directly or through a gradient to resolve the particles of interest.

More recently, setups employing more than one column in series have been developed. The advantage compared to single column setups is that the resin material can be fully loaded, since non-binding product is directly passed on to a consecutive column with fresh column material. The resin costs per amount of produced product can thus be drastically reduced. Since one column can always be eluted and regenerated while the other column is loaded, already two columns are sufficient to make full use of the advantages. Additional columns can give additional flexibility for elution and regeneration times, at the cost of additional equipment and resin costs.

Specific Uses

Affinity chromatography can be used in a number of applications, including nucleic acid purification, protein purification from cell free extracts, and purification from blood.

By using affinity chromatography, one can separate proteins that bind a certain fragment from proteins that do not bind that specific fragment. Because this technique of purification relies on the biological properties of the protein needed, it is a useful technique and proteins can be purified many folds in one step.

Various Affinity Media

Many different affinity media exist for a variety of possible uses. Briefly, they are (generalized):

- Activated/Functionalized – Works as a functional spacer, support matrix, and eliminates handling of toxic reagents.

- Amino Acid – Used with a variety of serum proteins, proteins, peptides, and enzymes, as well as rRNA and dsDNA.

- Avidin Biotin – Used in the purification process of biotin/avidin and their derivatives.

- Carbohydrate Bonding – Most often used with glycoproteins or any other carbohydrate-containing substance.

- Carbohydrate – Used with lectins, glycoproteins, or any other carbohydrate metabolite protein.

- Dye Ligand – This media is nonspecific, but mimics biological substrates and proteins.

- Glutathione – Useful for separation of GST tagged recombinant proteins.

- Heparin – This media is a generalized affinity ligand, and it is most useful for separation of plasma coagulation proteins, along with nucleic acid enzymes and lipases.

- Hydrophobic Interaction – Most commonly used to target free carboxyl groups and proteins.

- Immunoaffinity – Detailed below, this method utilizes antigens' and antibodies' high specificity to separate.

- Immobilized Metal Affinity Chromatography – Detailed further below, this method uses interactions between metal ions and proteins (usually specially tagged) to separate.

- Nucleotide/Coenzyme – Works to separate dehydrogenases, kinases, and transaminases.

- Nucleic Acid – Functions to trap mRNA, DNA, rRNA, and other nucleic acids/oligonucleotides.

- Protein A/G – This method is used to purify immunoglobulins.

- Speciality – Designed for a specific class or type of protein/coenzyme, this type of media will only work to separate a specific protein or coenzyme.

Immunoaffinity

Another use for the procedure is the affinity purification of antibodies from blood serum. If serum is known to contain antibodies against a specific antigen (for example if the serum comes from an organism immunized against the antigen concerned) then it can be used for the affinity purification of that antigen. This is also known as Immunoaffinity Chromatography. For example, if an organism is immunised against a GST-fusion protein it will produce antibodies against the fusion-protein, and possibly antibodies against the GST tag as well. The protein can then be covalently coupled to a solid support such as agarose and used as an affinity ligand in purifications of antibody from immune serum.

For thoroughness the GST protein and the GST-fusion protein can each be coupled separately. The serum is initially allowed to bind to the GST affinity matrix. This will remove antibodies against the GST part of the fusion protein. The serum is then separated from the solid support and allowed to bind to the GST-fusion protein matrix. This allows any antibodies that recognize the antigen to be captured on the solid support. Elution of the antibodies of interest is most often achieved using a low pH buffer such as glycine pH 2.8. The eluate is collected into a neutral tris or phosphate buffer, to neutralize the low pH elution buffer and halt any degradation of the antibody's activity. This is a nice example as affinity purification is used to purify the initial GST-fusion protein, to remove the undesirable anti-GST antibodies from the serum and to purify the target antibody.

A simplified strategy is often employed to purify antibodies generated against peptide antigens. When the peptide antigens are produced synthetically, a terminal cysteine residue is added at either the N- or C-terminus of the peptide. This cysteine residue contains a sulfhydryl functional group which allows the peptide to be easily conjugated to a carrier protein (e.g. Keyhole limpet hemocyanin (KLH)). The same cysteine-containing peptide is also immobilized onto an agarose resin through the cysteine residue and is then used to purify the antibody.

Most monoclonal antibodies have been purified using affinity chromatography based on immunoglobulin-specific Protein A or Protein G, derived from bacteria.

Immobilized Metal Ion Affinity Chromatography

Immobilized metal ion affinity chromatography (IMAC) is based on the specific coordinate covalent bond of amino acids, particularly histidine, to metals. This technique works by allowing proteins with an affinity for metal ions to be retained in a column containing immobilized metal ions, such as cobalt, nickel, copper for the purification of histidine containing proteins or peptides, iron, zinc or gallium for the purification of phosphorylated proteins or peptides. Many naturally occurring proteins do not have an affinity for metal ions, therefore recombinant DNA technology can be used to introduce such a protein tag into the relevant gene. Methods used to elute the protein of interest include changing the pH, or adding a competitive molecule, such as imidazole.

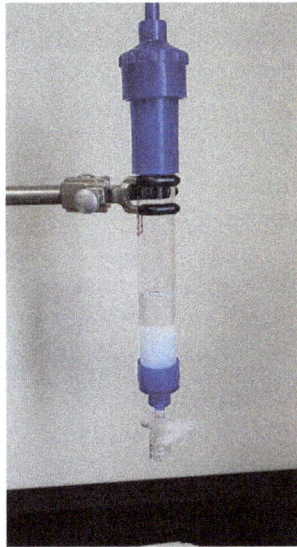

A chromatography column containing nickel-agarose beads used
for purification of proteins with histidine tags

Recombinant Proteins

Possibly the most common use of affinity chromatography is for the purification of recombinant proteins. Proteins with a known affinity are protein tagged in order to aid their purification. The protein may have been genetically modified so as to allow it to be selected for affinity binding; this is known as a fusion protein. Tags include gluta-thione-S-transferase (GST), hexahistidine (His), and maltose binding protein (MBP). Histidine tags have an affinity for nickel or cobalt ions which have been immobilized by forming coordinate covalent bonds with a chelator incorporated in the stationary phase. For elution, an excess amount of a compound able to act as a metal ion ligand, such as imidazole, is used. GST has an affinity for glutathione which is commercial-ly available immobilized as glutathione agarose. During elution, excess glutathione is used to displace the tagged protein.

Lectins

Lectin affinity chromatography is a form of affinity chromatography where lectins are used to separate components within the sample. Lectins, such as Concanavalin A are proteins which can bind specific alpha-D-mannose and alpha-D-glucose carbohydrate molecules. Another example of a lectin is wheat germ agglutinin which binds D-N-ace-tyl-glucosamine.The most common application is to separate glycoproteins from non-glycosylated proteins, or one glycoform from another glycoform.

Specialty

Another use for affinity chromatography is the purification of specific proteins using a gel matrix that is unique to a specific protein. For example, the purification of E.Coli-B-Galacto-sidase is accomplished by affinity chromatography using P-Aminobenyl-1-Thio-B-D-Ga-lactopyranosyl Agarose as the affinity matrix. P-Aminobenyl-1-Thio-B-D-Galactopyra-nosyl Agarose is used as the affinity matrix because it contains a galactopyranosyl group, which serves as a good substrate analog for E.Coli-B-Galactosidase. This property allows the enzyme to bind to the stationary phase of the affinity matrix and is eluted by adding increasing concentrations of salt to the column.

Boronate Affinity Chromatography consists of using boronic acid or boronates to elute and quantify amounts of glycoproteins. Clinical adaptations have applied this type of chromatography for use in determining long term assessment of diabetic patients through analyzation of their glycohemoglobin.

Immunoprecipitation

Immunoprecipitation (IP) is the technique of precipitating a protein antigen out of solution using an antibody that specifically binds to that particular protein. This pro-cess can be used to isolate and concentrate a particular protein from a sample con-taining many thousands of different proteins. Immunoprecipitation requires that the antibody be coupled to a solid substrate at some point in the procedure.

Types of Immunoprecipitation

Individual Protein Immunoprecipitation (IP)

Involves using an antibody that is specific for a known protein to isolate that particular protein out of a solution containing many different proteins. These solutions will often be in the form of a crude lysate of a plant or animal tissue. Other sample types could be body fluids or other samples of biological origin.

Protein Complex Immunoprecipitation (Co-IP)

Immunoprecipitation of intact protein complexes (i.e. antigen along with any proteins or ligands that are bound to it) is known as co-immunoprecipitation (Co-IP). Co-IP works by selecting an antibody that targets a known protein that is believed to be a member of a larger complex of proteins. By targeting this *known* member with an antibody it may become possible to pull the entire protein complex out of solution and thereby identify *unknown* members of the complex.

This works when the proteins involved in the complex bind to each other tightly, making it possible to pull multiple members of the complex out of solution by latching onto one member with an antibody. This concept of pulling protein complexes out of solution is sometimes referred to as a "pull-down". Co-IP is a powerful technique that is used regularly by molecular biologists to analyze protein–protein interactions.

- A particular antibody often selects for a subpopulation of its target protein that has the epitope exposed, thus failing to identify any proteins in complexes that hide the epitope. This can be seen in that it is rarely possible to precipitate even half of a given protein from a sample with a single antibody, even when a large excess of antibody is used.

- The first round that were not identified in the previous experiment. As successive rounds of targeting and immunoprecipitations take place, the number of identified proteins may continue to grow. The identified proteins may not ever exist in a single complex at a given time, but may instead represent a network of proteins interacting with one another at different times for different purposes.

- Repeating the experiment by targeting different members of the protein complex allows the researcher to double-check the result. Each round of pull-downs should result in the recovery of both the original known protein as well as other previously identified members of the complex (and even new additional members). By repeating the immunoprecipitation in this way, the researcher verifies that each identified member of the protein complex was a valid identification. If a particular protein can only be recovered by targeting one of the known members but not by targeting other of the known members then that protein's status as a member of the complex may be subject to question.

Chromatin Immunoprecipitation (ChIP)

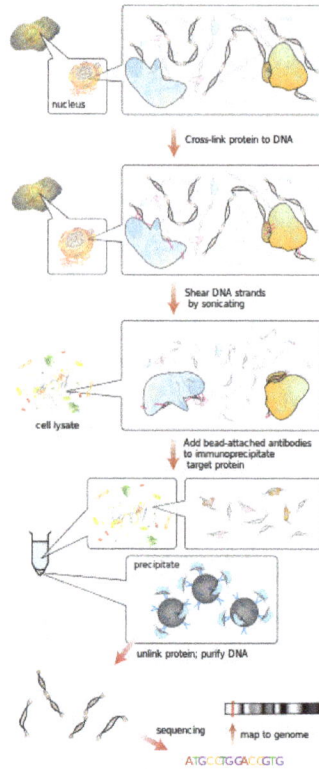

ChIP-sequencing workflow

Chromatin immunoprecipitation (ChIP) is a method used to determine the location of DNA binding sites on the genome for a particular protein of interest. This technique gives a picture of the protein–DNA interactions that occur inside the nucleus of living cells or tissues. The *in vivo* nature of this method is in contrast to other approaches traditionally employed to answer the same questions.

The principle underpinning this assay is that DNA-binding proteins (including transcription factors and histones) in living cells can be cross-linked to the DNA that they are binding. By using an antibody that is specific to a putative DNA binding protein, one can immunoprecipitate the protein–DNA complex out of cellular lysates. The crosslinking is often accomplished by applying formaldehyde to the cells (or tissue), although it is sometimes advantageous to use a more defined and consistent crosslinker such as DTBP. Following crosslinking, the cells are lysed and the DNA is broken into pieces 0.2–1.0 kb in length by sonication. At this point the immunoprecipitation is performed resulting in the purification of protein–DNA complexes. The purified protein–DNA complexes are then heated to reverse the formaldehyde cross-linking of the protein and DNA complexes, allowing the DNA to be separated from the proteins. The identity and quantity of the DNA fragments isolated can then be determined by PCR. The limitation of performing PCR on the isolated fragments is that one must have an idea which genomic region is being targeted in order to generate the correct PCR primers. Sometimes this limitation circumvented simply by cloning the isolated genomic DNA into a plasmid vector and then using primers that are specific to the cloning region of that vector. Alternatively, when one wants to find where the protein binds on a genome-wide scale, ChIP-Sequencing is used and has recently emerged as a standard technology that can localize protein binding sites in a high-throughput, cost-effective fashion, allowing also for the characterization of the cistrome. Previously, DNA microarray was also used (ChIP-on-chip or ChIP-chip).

RNP Immunoprecipitation (RIP)

Similar to chromatin immunoprecipitation (ChIP) outlined above, but rather than targeting DNA binding proteins as in ChIP, an RNP immunoprecipitation targets ribonucleoproteins (RNPs). Live cells are first lysed and then the target protein and associated RNA are immunoprecipitated using an antibody targeting the protein of interest. The purified RNA-protein complexes can be separated by performing an RNA extraction and the identity of the RNA can be determined by cDNA sequencing or RT-PCR. Some variants of RIP, such as PAR-CLIP include cross-linking steps, which then require less careful lysis conditions.

Tagged Proteins

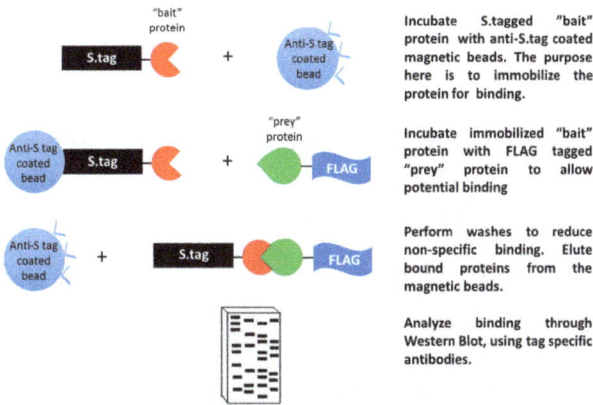

Pull down assay using tagged proteins

One of the major technical hurdles with immunoprecipitation is the great difficulty in generating an antibody that specifically targets a single known protein. To get around this obstacle, many groups will engineer tags onto either the C- or N- terminal end of the protein of interest. The advantage here is that the same tag can be used time and again on many different proteins and the researcher can use the same antibody each time. The advantages with using tagged proteins are so great that this technique has become commonplace for all types of immunoprecipitation including all of the types of IP detailed above. Examples of tags in use are the Green Fluorescent Protein (GFP) tag, Glutathione-S-transferase (GST) tag and the FLAG-tag tag. While the use of a tag to enable pull-downs is convenient, it raises some concerns regarding biological relevance because the tag itself may either obscure native interactions or introduce new and unnatural interactions.

Methods

The two general methods for immunoprecipitation are the direct capture method and the indirect capture method.

Direct

Antibodies that are specific for a particular protein (or group of proteins) are immobilized on a solid-phase substrate such as superparamagneticmicrobeads or on microscopic agarose (non-magnetic) beads. The beads with bound antibodies are then added to the protein mixture, and the proteins that are targeted by the antibodies are captured onto the beads via the antibodies; in other words, they become immunoprecipitated.

Indirect

Antibodies that are specific for a particular protein, or a group of proteins, are added directly to the mixture of protein. The antibodies have not been attached to a solid-phase support yet. The antibodies are free to float around the protein mixture and bind their targets. As time passes, the beads coated in protein A/G are added to the mixture of antibody and protein. At this point, the antibodies, which are now bound to their targets, will stick to the beads.

From this point on, the direct and indirect protocols converge because the samples now have the same ingredients. Both methods gives the same end-result with the protein or protein complexes bound to the antibodies which themselves are immobilized onto the beads.

Selection

An indirect approach is sometimes preferred when the concentration of the protein target is low or when the specific affinity of the antibody for the protein is weak. The indirect method is also used when the binding kinetics of the antibody to the protein is slow for a variety of reasons. In most situations, the direct method is the default, and the preferred, choice.

Technological Advances

Agarose

Historically the solid-phase support for immunoprecipitation used by the majority of scientists has been highly-porous agarose beads (also known as agarose resins or slurries). The advantage of this technology is a very high potential binding capacity, as virtually the entire sponge-like structure of the agarose particle (50 to 150μm in size) is available for binding antibodies (which will in turn bind the target proteins) and the use of standard laboratory equipment for all aspects of the IP protocol without the need for any specialized equipment. The advantage of an extremely high binding capacity must be carefully balanced with the quantity of antibody that the researcher is prepared to use to coat the agarose beads. Because antibodies can be a cost-limiting factor, it is best to calculate backward *from* the amount of protein that needs to be captured (depending upon the analysis to be performed downstream), *to* the amount of antibody that is required to bind that quantity of protein (with a small excess added in order to account for inefficiencies of the system), and back still further *to* the quantity of agarose that is needed to bind that particular quantity of antibody. In cases where antibody saturation is not required, this technology is unmatched in its ability to capture extremely large quantities of captured target proteins. The caveat here is that the *"high capacity advantage"* can become a *"high capacity disadvantage"* that is manifested when the enormous binding capacity of the sepharose/agarose beads is not completely saturated with antibodies. It often happens that the amount of antibody available to the researcher for their immunoprecipitation experiment is less than sufficient to saturate the agarose beads to be used in the immunoprecipitation. In these cases the researcher can end up with agarose particles that are only partially coated with antibodies, and the portion of the binding capacity of the agarose beads that is not coated with antibody is then free to bind anything that will stick, resulting in an elevated background signal due to non-specific binding of lysate components to the beads, which can make data interpretation difficult. While some may argue that for these reasons it is prudent to match the quantity of agarose (in terms of binding capacity) to the quantity of antibody that one wishes to be bound for the immunoprecipitation, a simple way to reduce the issue of non-specific binding to agarose beads and increase specificity is to preclear the lysate, which for any immunoprecipitation is highly recommended.

Preclearing

Lysates are complex mixtures of proteins, lipids, carbohydrates and nucleic acids, and one must assume that some amount of non-specific binding to the IP antibody, Protein A/G or the beaded support will occur and negatively affect the detection of the immunoprecipitated target(s). In most cases, *preclearing* the lysate at the start of each immunoprecipitation experiment is a way to remove potentially reactive components from the cell lysate prior to the immunoprecipitation to prevent the non-specific binding of these components to the IP beads or antibody. The basic preclearing procedure is described below, wherein the lysate is incubated with beads alone, which are then removed and discarded prior to the immunoprecipitation This approach, though, does not account for non-specific binding to the IP antibody, which can be considerable. Therefore, an alternative method of preclearing is to incubate the protein mixture with exactly the same components that will be used in the immunoprecipitation, except that a non-target, irrelevant antibody of the same antibody subclass as the IP antibody is used instead of the IP antibody itself. This approach attempts to use as close to the exact IP conditions and components as the actual immunoprecipitation to remove any non-specific cell constituent without capturing the target protein (unless, of course, the target protein non-specifically binds to some other IP component, which should be properly controlled for by analyzing the discarded beads used to preclear the lysate). The target protein can then be immunoprecipitated with the reduced risk of non-specific binding interfering with data interpretation.

Superparamagnetic Beads

While the vast majority of immunoprecipitations are performed with agarose beads, the use of superparamagnetic beads for immunoprecipitation is a much newer approach that is only recently gaining in popularity as an alternative to agarose beads for IP applications. Unlike agarose, magnetic beads are solid and can be spherical, depending on the type of bead, and antibody binding is limited to the surface of each bead. While these beads do not have the advantage of a porous center to increase the binding capacity, magnetic beads are significantly smaller than agarose beads (1 to 4μm), and the greater number of magnetic beads per volume than agarose beads collectively gives magnetic beads an effective surface area-to-volume ratio for optimum antibody binding.

Commercially available magnetic beads can be separated based by size uniformity into monodisperse and polydisperse beads. Monodisperse beads, also called microbeads, exhibit exact uniformity, and therefore all beads exhibit identical physical characteristics, including the binding capacity and the level of attraction to magnets. Polydisperse beads, while similar in size to monodisperse beads, show a wide range in size variability (1 to 4μm) that can influence their binding capacity and magnetic capture. Although both types of beads are commercially available for immunoprecipitation applications, the higher quality monodispersesuperparamagnetic beads are more ideal for automatic protocols because of their consistent size, shape and performance. Monodisperse and polydispersesuperparamagnetic beads are offered by many companies, including Invitrogen, Thermo Scientific, and Millipore.

Agarose vs. Magnetic Beads

Proponents of magnetic beads claim that the beads exhibit a faster rate of protein binding over agarose beads for immunoprecipitation applications, although standard agarose bead-based immunoprecipitations have been performed in 1 hour. Claims have also been made that magnetic beads are better for immunoprecipitating extremely large protein complexes because of the complete lack of an upper size limit for such complexes, although there is no unbiased evidence stating this claim. The nature of magnetic bead technology does result in less sample handling due to the reduced physical stress on samples of magnetic separation versus repeated centrifugation when using agarose, which may contribute greatly to increasing the yield of labile (fragile) protein complexes. Additional factors, though, such as the binding capacity, cost of the reagent, the requirement of extra equipment and the capability to automate IP processes should be considered in the selection of an immunoprecipitation support.

Binding Capacity

Proponents of both agarose and magnetic beads can argue whether the vast difference in the binding capacities of the two beads favors one particular type of bead. In a bead-to-bead comparison, agarose beads have significantly greater surface area and therefore a greater binding capacity than magnetic beads due to the large bead size and sponge-like structure. But the variable pore size of the agarose causes a potential upper size limit that may affect the binding of extremely large proteins or protein complexes to internal binding sites, and therefore magnetic beads may be better suited for immunoprecipitating large proteins or protein complexes than agarose beads, although there is a lack of independent comparative evidence that proves either case.

Some argue that the significantly greater binding capacity of agarose beads may be a disadvantage because of the larger capacity of non-specific binding. Others may argue for the use of magnetic beads because of the greater quantity of antibody required to saturate the total binding capacity of agarose beads, which would obviously be an economical disadvantage of using agarose. While these arguments are correct outside the context of their practical use, these lines of reasoning ignore two key aspects of the principle of immunoprecipitation that demonstrates that the decision to use agarose or magnetic beads is not simply determined by binding capacity.

First, non-specific binding is not limited to the antibody-binding sites on the immobilized support; any surface of the antibody or component of the immunoprecipitation reaction can bind to nonspecific lysate constituents, and therefore nonspecific binding will still occur even when completely saturated beads are used. This is why it is important to preclear the sample before the immunoprecipitation is performed.

Second, the ability to capture the target protein is directly dependent upon the amount of immobilized antibody used, and therefore, in a side-by-side comparison of agarose and magnetic bead immunoprecipitation, the most protein that either support can capture is limited by the amount of antibody added.

Cost

The price of using either type of support is a key determining factor in using agarose or magnetic beads for immunoprecipitation applications. A typical first-glance calculation on the cost of magnetic beads compared to sepharose beads may make the sepharose beads appear less expensive. But magnetic beads may be competitively priced compared to agarose for analytical-scale immunoprecipitations depending on the IP method used and the volume of beads required per IP reaction.

Using the traditional batch method of immunoprecipitation as listed below, where all components are added to a tube during the IP reaction, the physical handling characteristics of agarose beads necessitate a minimum quantity of beads for each IP experiment (typically in the range of 25 to 50μl beads per IP). This is because sepharose beads must be concentrated at the bottom of the tube by centrifugation and the supernatant removed after each incubation, wash, etc. This imposes absolute physical limitations on the process, as pellets of agarose beads less than 25 to 50μl are difficult if not impossible to visually identify at the bottom of the tube. With magnetic beads, there is no minimum quantity of beads required due to magnetic handling, and therefore, depending on the target antigen and IP antibody, it is possible to use considerably less magnetic beads.

Conversely, spin columns may be employed instead of normal microfuge tubes to significantly reduce the amount of agarose beads required per reaction. Spin columns contain a filter that allows all IP components except the beads to flow through using a brief centrifugation and therefore provide a method to use significantly less agarose beads with minimal loss.

Equipment

As mentioned above, only standard laboratory equipment is required for the use of agarose beads in immunoprecipitation applications, while high-power magnets are required for magnetic bead-based IP reactions. While the magnetic capture equipment may be cost-prohibitive, the rapid completion of immunoprecipitations using magnetic beads may be a financially beneficial approach when grants are due, because a 30-minute protocol with magnetic beads compared to overnight incubation at 4 °C with agarose beads may result in more data generated in a shorter length of time.

Automation

An added benefit of using magnetic beads is that automated immunoprecipitation devices are becoming more readily available. These devices not only reduce the amount of work and time to perform an IP, but they can also be used for high-throughput applications.

While clear benefits of using magnetic beads include the increased reaction speed, more gentle sample handling and the potential for automation, the choice of using agarose or magnetic beads based on the binding capacity of the support medium and the cost of the product may depend on the protein of interest and the IP method used. As with all assays, empirical testing is required to determine which method is optimal for a given application.

Protocol

Background

Once the solid substrate bead technology has been chosen, antibodies are coupled to the beads and the antibody-coated-beads can be added to the heterogeneous protein sample (e.g. homogenized tissue). At this point, antibodies that are immobilized to the beads will bind to the proteins that they specifically recognize. Once this has occurred the immunoprecipitation portion of the protocol is actually complete, as the specific proteins of interest are bound to the antibodies that are themselves immobilized to the beads. Separation of the immunocomplexes from the lysate is an extremely important series of steps, because the protein(s) must remain bound to each other (in the case of co-IP) and bound to the antibody during the wash steps to remove non-bound proteins and reduce background.

When working with agarose beads, the beads must be pelleted out of the sample by briefly spinning in a centrifuge with forces between 600–3,000 x g (times the standard gravitational force). This step may be performed in a standard microcentrifuge tube, but for faster separation, greater consistency and higher recoveries, the process is often performed in small spin columns with a pore size that allows liquid, but not agarose beads, to pass through. After centrifugation, the agarose beads will form a very loose fluffy pellet at the bottom of the tube. The supernatant containing contaminants can be carefully removed so as not to disturb the beads. The wash buffer can then be added to the beads and after mixing, the beads are again separated by centrifugation.

With superparamagnetic beads, the sample is placed in a magnetic field so that the beads can collect on the side of the tube. This procedure is generally complete in approximately 30 seconds, and the remaining (unwanted) liquid is pipetted away. Washes are accomplished by resuspending the beads (off the magnet) with the washing solution and then concentrating the beads back on the tube wall (by placing the tube back on the magnet). The washing is generally repeated several times to ensure adequate removal of contaminants. If the superparamagnetic beads are homogeneous in size and the magnet has been designed properly, the beads will concentrate uniformly on the side of the tube and the washing solution can be easily and completely removed.

After washing, the precipitated protein(s) are eluted and analyzed by gel electrophoresis, mass spectrometry, western blotting, or any number of other methods for identifying constituents in the complex. Protocol times for immunoprecipitation vary greatly due to a variety of factors, with protocol times increasing with the number of washes necessary or with the slower reaction kinetics of porous agarose beads.

Steps

1.   Lyse cells and prepare sample for immunoprecipitation.

2.   Pre-clear the sample by passing the sample over beads alone or bound to an irrelevant antibody to soak up any proteins that non-specifically bind to the IP components.

3. Incubate solution with antibody against the protein of interest. Antibody can be attached to solid support before this step (direct method) or after this step (indirect method). Continue the incubation to allow antibody-antigen complexes to form.

4. Precipitate the complex of interest, removing it from bulk solution.

5. Wash precipitated complex several times. Spin each time between washes when using agarose beads or place tube on magnet when using superparamagnetic beads and then remove the supernatant. After the final wash, remove as much supernatant as possible.

6. Elute proteins from the solid support using low-pH or SDS sample loading buffer.

7. Analyze complexes or antigens of interest. This can be done in a variety of ways:

   1. SDS-PAGE (sodium dodecyl sulfate-polyacrylamidegel electrophoresis) followed by gel staining.

   2. SDS-PAGE followed by: gel staining, cutting out individual stained protein bands, and sequencing the proteins in the bands by MALDI-Mass Spectrometry

   3. Transfer and Western Blot using another antibody for proteins that were interacting with the antigen followed by chemiluminesent visualization.

Co-Immunoprecipitation

This technique works by using an antibody which immunoprecipitate the respective antigen and with it co - immunoprecipitate any interacting protein. The protein molecules involved in the complex formation must be bound tightly to each other so that on being pulled, entire complex can be precipitated out. The method checks the interaction in the protein complex in their natural conformation. However one disadvantage is that sometimes concentration of the antigen gets lower thus unable to be detect ed effeciently.

Important steps involved in Co-immunoprecipitation technique

1. Antibody- Antigen selection

It is very important to select specific antigen and antibody (for precipitating the antigen bounded with protein of interest), for easy and specific precipitation. Sometimes it happens that the antibody precipitate a non specific protein other than desired antigenic protein.

2. Cellular lysis procedure.

Care should also be taken during the cellular lysis procedure as not to disturb the proteins complex. The lysis procedure should be mild.

Tandem Affinity Purification

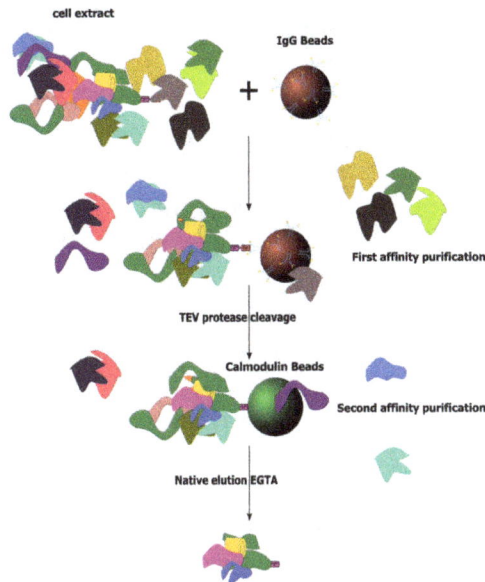

cell extract

IgG Beads

+

First affinity purification

TEV protease cleavage

Calmodulin Beads

Second affinity purification

Native elution EGTA

Tandem affinity purification (TAP) is a purification technique for studying protein–protein interactions. It involves creating a fusion protein with a designed piece, the TAP tag, on the end.

In the original version of the technique, the protein of interest with the TAP tag first binds to beads coated with IgG, the TAP tag is then broken apart by an enzyme, and finally a different part of the TAP tag binds reversibly to beads of a different type. After the protein of interest has been washed through two affinity columns, it can be examined for binding partners.

The original TAP method involves the fusion of the TAP tag to the C-terminus of the protein under study. The TAP tag consists of calmodulin binding peptide (CBP) from the N-terminal, followed by tobacco etch virus protease (TEV protease) cleavage site and Protein A, which binds tightly to IgG. The relative order of the modules of the tag is important because Protein A needs to be at the extreme end of the fusion protein so that the entire complex can be retrieved using an IgG matrix.

Many other tag combinations have been proposed since the TAP principle was first published.

Variant Tags

This tag is also known as the C-terminal TAP tag because an N-terminal version is also available. However, the method to be described assumes the use of a C-terminal tag, although the principle behind the method is still the same.

History

TAP tagging was invented by a research team working in the European Molecular Biology Laboratory at late 1990s (Rigaut et al., 1999, Puig et al.,2001) and proposed as a new tool for proteome exploration. It was used by the team to characterize several protein complexes (Rigaut et al., 1999, Caspary et al. 1999, Bouveret et al., 2000, Puig et al., 2001). The first large-scale application of this technique was in 2002, in which the research team worked in collaboration with scientists of the proteomics company Cellzome to develop a visual map of the interaction of more than 230 multi-protein complexes in a yeast cell by systematically tagging the TAP tag to each protein. The first successful report of using TAP tag technology in plants came in 2004 (Rohila et al., 2004,)

Process

There are a few methods in which the fusion protein can be introduced into the host. If the host is yeast, then one of the methods may be the use of plasmids that will eventually translate the fusion protein within the host. Whichever method that is being used, it is preferable to maintain expression of the fusion protein as close as possible to its natural level.

Once the fusion protein is translated within the host, the new protein at one end of the fusion protein would be able to interact with other proteins. Subsequently, the fusion protein is retrieved from the host by breaking the cells and retrieving the fusion protein through affinity selection, together with the other constituents attached to the new protein, by means of an IgG matrix.

After washing, TEV protease is introduced to elute the bound material at the TEV protease cleavage site. This eluate is then incubated with calmodulin-coated beads in the presence of calcium. This second affinity step is required to remove the TEV protease as well as traces of contaminants remaining after the first affinity step. After washing, the eluate is then released with ethylene glycol tetraacetic acid (EGTA).

The native elution, consisting of the new protein and its interacting protein partners as well as CBP, can now be analyzed by sodium dodecyl sulfate polyacrylamide gel electrophoresis (SDS-PAGE) or be identified by mass spectrometry.

Advantages

An advantage of this method is that there can be real determination of protein partners quantitatively in vivo without prior knowledge of complex composition. It is also simple to execute and often provides high yield. One of the obstacles of studying protein protein interaction is the contamination of the target protein especially when we don't have any prior knowledge of it. TAP offers an effective, and highly specific means to purify target protein. After 2 successive affinity purifications, the chance for contaminants to be retained in the eluate reduces significantly.

Disadvantages

However, there is also the possibility that a tag added to a protein might obscure binding of the new protein to its interacting partners. In addition, the tag may also affect protein expression levels. On the other hand, the tag may also not be sufficiently exposed to the affinity beads, hence skewing the results.

There may also be a possibility of a cleavage of the proteins by the TEV protease, although this is unlikely to be frequent given the high specificity of the TEV protease.

Suitability

As this method involves at least 2 rounds of washing, it may not be suitable for screening transient protein interactions, unlike the yeast two-hybrid method or *in vivo* crosslinking with photo-reactive amino acid analogs. However, it is a good method for testing stable protein interactions and allows various degrees of investigation by controlling the number of times the protein complex is purified.

Applications

In 2002, the TAP tag was first used with mass spectrometry in a large-scale approach to systematically analyse the proteomics of yeast by characterizing multiprotein complexes. The study revealed 491 complexes, 257 of them wholly new. The rest were familiar from other research, but now virtually all of them were found to have new components. They drew up a map relating all the protein components functionally in a complex network.

Many other proteomic analyses also involve the use of TAP tag. A research by EMBO (Dziembowski, 2004) identified a new complex required for nuclear pre-mRNA retention and splicing. They have purified a novel trimeric complex composed of 3 other subunits (Snu17p, Bud13p and Pml1p) and find that these subunits are not essential for viability but required for efficient splicing (removal of introns) of pre-mRNA. In 2006, *Fleischer et al.* systematically identified proteins associated with eukaryotic ribosomal complexes. They used multifaceted mass spectrometry proteomic screens to identify yeast ribosomal complexes and then used TAP tagging to functionally link up all these proteins.

Other Epitope-tag Combinations

The principle of tandem-affinity purification of multiprotein complexes is not limited to the combination of CBP and Protein A tags used in the original work by Rigaut et al. (1999). For example, the combination of FLAG- and HA-tags has been used since 2000 by the group of Nakatani to purify numerous protein complexes from mammalian cells. Many other tag combinations have been proposed since the TAP principle was published.

Protein Mass Spectrometry

A mass spectrometer used for high throughput protein analysis.

Protein mass spectrometry refers to the application of mass spectrometry to the study of proteins. Mass spectrometry is an important method for the accurate mass determination and characterization of proteins, and a variety of methods and instrumentations have been developed for its many uses. Its applications include the identification of proteins and their post-translational modifications, the elucidation of protein complexes, their subunits and functional interactions, as well as the global measurement of proteins in proteomics. It can also be used to localize proteins to the various organelles, and determine the interactions between different proteins as well as with membrane lipids.

The two primary methods used for the ionization of protein in mass spectrometry are electrospray ionization (ESI) and matrix-assisted laser desorption/ionization (MALDI). These ionization techniques are used in conjunction with mass analyzers such as tandem mass spectrometry. In general, the protein are analyzed either in a "top-down" approach in which proteins are analyzed intact, or a "bottom-up" approach in which protein are first digested into fragments. An intermediate "middle-down" approach in which larger peptide fragments are analyzed may also sometimes be used.

History

The application of mass spectrometry to study proteins became popularized in the 1980s after the development of MALDI and ESI. These ionization techniques have played a significant role in the characterization of proteins. (MALDI) Matrix-assisted laser desorption ionization was coined in the late 80's by Franz Hillenkamp and Michael Karas. Hillenkamp, Karas and their fellow researchers were able to ionize the amino acid alanine by mixing it with the amino acid tryptophan and irradiated with a pulse 266 nm laser. Though important, the breakthrough did not come until 1987. In 1987, Koichi Tanaka used the "ultra fine metal plus liquid matrix method" and ionized biomolecules the size of 34,472 Da protein carboxypeptidase-A.

In 1968, Malcolm Dole reported the first use of electrospray ionization with mass spectrometry. Around the same time MALDI became popularized, John Bennett Fenn was cited for the development of electrospray ionization. Koichi Tanaka received the 2002 Nobel Prize in Chemistry alongside John Fenn, and Kurt Wüthrich "for the development of methods for identification and structure analyses of biological macromolecules." These ionization methods have greatly facilitated the study of proteins by mass spectrometry. Consequently, protein mass spectrometry now plays a leading role in protein characterization.

Methods and Approaches

Techniques

Mass spectrometry of proteins requires that the proteins in solution or solid state be turned into an ionized form in the gas phase before they are injected and accelerated in an electric or magnetic field for analysis. The two primary methods for ionization of proteins are electrospray ionization (ESI) and matrix-assisted laser desorption/ionization (MALDI). In electrospray, the ions are created from proteins in solution, and it allows fragile molecules to be ionized intact, sometimes preserving non-covalent interactions. In MALDI, the proteins are embedded within a matrix normally in a solid form, and ions are created by pulses of laser light. Electrospray produces more multiply-charged ions than MALDI, allowing for measurement of high mass protein and better fragmentation for identification, while MALDI is fast and less likely to be affected by contaminants, buffers and additives.

Whole-protein mass analysis is primarily conducted using either time-of-flight (TOF) MS, or Fourier transform ion cyclotron resonance (FT-ICR). These two types of instrument are preferable here because of their wide mass range, and in the case of FT-ICR, its high mass accuracy. Electrospray ionization of a protein often results in generation of multiple charged species of $800 < m/z < 2000$ and the resultant spectrum can be deconvoluted to determine the protein's average mass to within 50 ppm or better using TOF or ion-trap instruments.

Mass analysis of proteolytic peptides is a popular method of protein characterization, as cheaper instrument designs can be used for characterization. Additionally, sample preparation is easier once whole proteins have been digested into smaller peptide fragments. The most widely used instrument for peptide mass analysis are the MALDI-TOF instruments as they permit the acquisition of peptide mass fingerprints (PMFs) at high pace (1 PMF can be analyzed in approx. 10 sec). Multiple stage quadrupole-time-of-flight and the quadrupole ion trap also find use in this application.

Chromatography trace and MS/MS spectra of a peptide.

Tandem mass spectrometry (MS/MS) is used to measure fragmentation spectra and identify proteins at high speed and accuracy. Collision-induced dissociation is used in mainstream applications to generate a set of fragments from a specific peptide ion. The fragmentation process primarily gives rise to cleavage products that break along peptide bonds. Because of this simplicity in fragmentation, it is possible to use the observed fragment masses to match with a database of predicted masses for one of many given peptide sequences. Tandem MS of whole protein ions has been investigated recently using electron capture dissociation and has demonstrated extensive sequence information in principle but is not in common practice.

Approaches

In keeping with the performance and mass range of available mass spectrometers, two approaches are used for characterizing proteins. In the first, intact proteins are ionized by either of the two techniques described above, and then introduced to a mass analyzer. This approach is referred to as "top-down" strategy of protein analysis as it involves starting with the whole mass and then pulling it apart. The top-down approach however is mostly limited to low-throughput single-protein studies due to issues involved in handling whole proteins, their heterogeneity and the complexity of their analyses.

In the second approach, referred to as the "bottom-up" MS, proteins are enzymatically digested into smaller peptides using a protease such as trypsin. Subsequently, these peptides are introduced into the mass spectrometer and identified by peptide mass fingerprinting or tandem mass spectrometry. Hence, this approach uses identification at the peptide level to infer the existence of proteins pieced back together with *de novo* repeat detection. The smaller and more uniform fragments are easier to analyze than intact proteins and can be also determined with high accuracy, this "bottom-up" approach is therefore the preferred method of studies in proteomics. A further approach that is beginning to be useful is the intermediate "middle-down" approach in which proteolytic peptides larger than the typical tryptic peptides are analyzed.

Protein and Peptide Fractionation

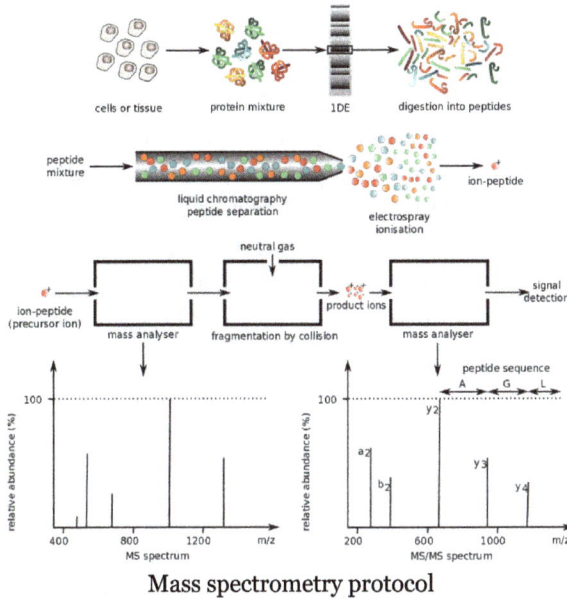

Mass spectrometry protocol

Proteins of interest are usually part of a complex mixture of multiple proteins and molecules, which co-exist in the biological medium. This presents two significant problems. First, the two ionization techniques used for large molecules only work well when the mixture contains roughly equal amounts of constituents, while in biological samples, different proteins tend to be present in widely differing amounts. If such a mixture is ionized using electrospray or MALDI, the more abundant species have a tendency to "drown" or suppress signals from less abundant ones. Second, mass spectrum from a complex mixture is very difficult to interpret due to the overwhelming number of mixture components. This is exacerbated by the fact that enzymatic digestion of a protein gives rise to a large number of peptide products.

In light of these problems, the methods of one- and two-dimensional gel electrophoresis and high performance liquid chromatography are widely used for separation of proteins. The first method fractionates whole proteins via two-dimensional gel electrophoresis. The first-dimension of 2D gel is isoelectric focusing (IEF). In this dimension, the protein is separated by its isoelectric point (pI) and the second-dimension is SDS-polyacrylamide gel electrophoresis (SDS-PAGE). This dimension separates the protein according to its molecular weight. Once this step is completed in-gel digestion occurs. In some situations, it may be necessary to combine both of these techniques. Gel spots identified on a 2D Gel are usually attributable to one protein. If the identity of the protein is desired, usually the method of in-gel digestion is applied, where the protein spot of interest is excised, and digested proteolytically. The peptide masses resulting from the digestion can be determined by mass spectrometry using peptide mass fingerprinting. If this information does not allow unequivocal identification of the protein, its peptides can be subject to tandem mass spectrometry for *de novo* sequencing. Small changes in mass and charge can be detected with 2D-PAGE. The disadvantages with this technique are its small dynamic range compared to other methods, some proteins are still difficult to separate due to their acidity, basicity, hydrophobicity, and size (too large or too small).

The second method, high performance liquid chromatography is used to fractionate peptides after enzymatic digestion. Characterization of protein mixtures using HPLC/MS is also called shotgun proteomics and MuDPIT (Multi-Dimensional Protein Identification Technology). A peptide mixture that results from digestion of a protein mixture is fractionated by one or two steps of liquid chromatography. The eluent from the chromatography stage can be either directly introduced to the mass spectrometer through electrospray ionization, or laid down on a series of small spots for later mass analysis using MALDI.

Applications

Protein Identification

There are two main ways MS is used to identify proteins. Peptide mass fingerprinting uses the masses of proteolytic peptides as input to a search of a database of predicted masses that would arise from digestion of a list of known proteins. If a protein sequence in the reference list gives rise to a significant number of predicted masses that match the experimental values, there is some evidence that this protein was present in the original sample. Purification steps therefore limit the throughput of the peptide mass fingerprinting approach. Peptide mass fingerprinting can be achieved with MS/MS.

MS is also the preferred method for the identification of post-translational modifications in proteins as it is more advantageous than other approaches such as the antibody-based methods.

De Novo (Peptide) Sequencing

De novo peptide sequencing for mass spectrometry is typically performed without prior knowledge of the amino acid sequence. It is the process of assigning amino acids from peptide fragment masses of a protein. *De novo* sequencing has proven successful for confirming and expanding upon results from database searches.

As *de novo* sequencing is based on mass and some amino acids have identical masses (e.g. leucine and isoleucine), accurate manual sequencing can be difficult. Therefore, it may be necessary to utilize a sequence homology search application to work in tandem between a database search and *de novo* sequencing to address this inherent limitation.

Database searching has the advantage of quickly identifying sequences, provided they have already been documented in a database. Other inherent limitations of database searching include sequence modifications/mutations (some database searches do not adequately account for alterations to the 'documented' sequence, thus can miss valuable information), the unknown (if a sequence is not documented, it will not be found), false positives, and incomplete and corrupted data.

An annotated peptide spectral library can also be used as a reference for protein/peptide identification. It offers the unique strength of reduced search space and increased specificity. The limitations include spectra not included in the library will not be identified, spectra collected from different types of mass spectrometers can have quite distinct features, and reference spectra in the library may contain noise peaks, which may lead to false positive identifications. A number of different algorithmic approaches have been described to identify peptides and proteins from tandem mass spectrometry (MS/MS), peptide *de novo* sequencing and sequence tag-based searching.

Protein Quantitation

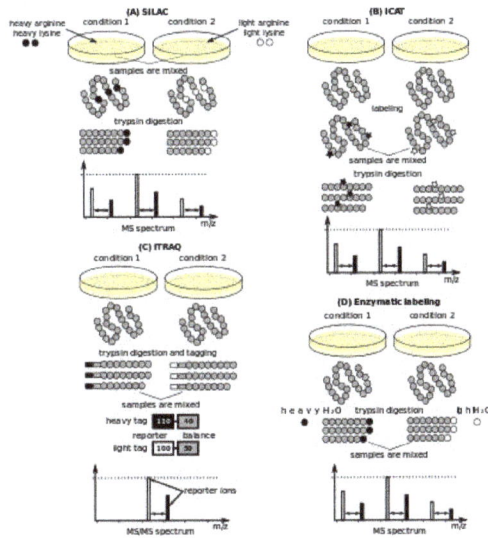

Quantitative Mass Spectrometry.

Several recent methods allow for the quantitation of proteins by mass spectrometry (quantitative proteomics). Typically, stable (e.g. non-radioactive) heavier isotopes of carbon (^{13}C) or nitrogen (^{15}N) are incorporated into one sample while the other one is labeled with corresponding light isotopes (e.g. ^{12}C and ^{14}N). The two samples are mixed before the analysis. Peptides derived from the different samples can be distinguished due to their mass difference. The ratio of their peak intensities corresponds to the relative abundance ratio of the peptides (and proteins). The most popular methods for isotope labeling are SILAC (stable isotope labeling by amino acids in cell culture), trypsin-catalyzed ^{18}O labeling, ICAT (isotope coded affinity tagging), iTRAQ (isobaric tags for relative and absolute quantitation). "Semi-quantitative" mass spectrometry can be performed without labeling of samples. Typically, this is done with MALDI analysis (in linear mode). The peak intensity, or the peak area, from individual molecules (typically proteins) is here correlated to the amount of protein in the sample. However, the individual signal depends on the primary structure of the protein, on the complexity of the sample, and on the settings of the instrument. Other types of "label-free" quantitative mass spectrometry, uses the spectral counts (or peptide counts) of digested proteins as a means for determining relative protein amounts.

Protein Structure Determination

Characteristics indicative of the 3-dimensional structure of proteins can be probed with mass spectrometry in various ways. By using chemical crosslinking to couple parts of the protein that are close in space, but far apart in sequence, information about the overall structure can be inferred. By following the exchange of amide protons with deuterium from the solvent, it is possible to probe the solvent accessibility of various parts of the protein.Hydrogen-deuterium exchange mass spectrometry has been used to study proteins and their conformations for over 20 years. This type of protein structural analysis can be suitable for proteins that are challenging for other structural methods. Another interesting avenue in protein structural studies is laser-induced covalent labeling. In this technique, solvent-exposed sites of the protein are modified by hydroxyl radicals. Its combination with rapid mixing has been used in protein folding studies.

Biomarkers

The FDA defines a biomarker as, "A characteristic that is objectively measured and evaluated as an indicator of normal biologic processes, pathogenic processes, or pharmacologic responses to a therapeutic intervention". It is hypothesized that mass spectrometry enables the discovery of candidates for biomarkers.

Proteogenomics

In what is now commonly referred to as proteogenomics, peptides identified with mass spectrometry are used for improving gene annotations (for example, gene start sites) and protein annotations. Parallel analysis of the genome and the proteome facilitates discovery of post-translational modifications and proteolytic events, especially when comparing multiple species.

Electrospray Ionization

Electrospray ionization (ESI) is a technique used in mass spectrometry to produce ions using an electrospray in which a high voltage is applied to a liquid to create an aerosol. It is especially useful in producing ions from macromolecules because it overcomes the propensity of these molecules to fragment when ionized. ESI is different from other atmospheric pressure ionization processes (e.g. matrix-assisted laser desorption/ionization (MALDI)) since it may produce multiple charged ions, effectively extending the mass range of the analyser to accommodate the kDa-MDa orders of magnitude observed in proteins and their associated polypeptide fragments.

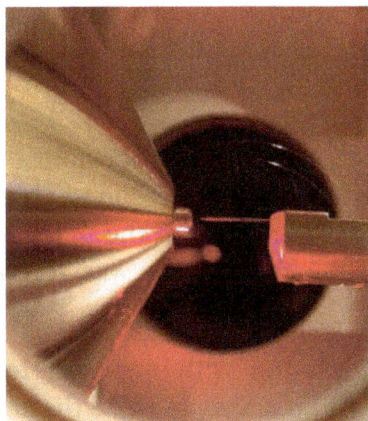

Electrospray (nanoSpray) ionization source

Mass spectrometry using ESI is called electrospray ionization mass spectrometry (ESI-MS) or, less commonly, electrospray mass spectrometry (ES-MS). ESI is a so-called 'soft ionization' technique, since there is very little fragmentation. This can be advantageous in the sense that the molecular ion (or more accurately a pseudo molecular ion) is always observed, however very little structural information can be gained from the simple mass spectrum obtained. This disadvantage can be overcome by coupling ESI with tandem mass spectrometry (ESI-MS/MS). Another important advantage of ESI is that solution-phase information can be retained into the gas-phase.

The electrospray ionization technique was first reported by Masamichi Yamashita and John Fenn in 1984. The development of electrospray ionization for the analysis of biological macromolecules was rewarded with the attribution of the Nobel Prize in Chemistry to John Bennett Fenn in 2002. One of the original instruments used by Dr. Fenn is on display at the Chemical Heritage Foundation in Philadelphia, Pennsylvania.

History

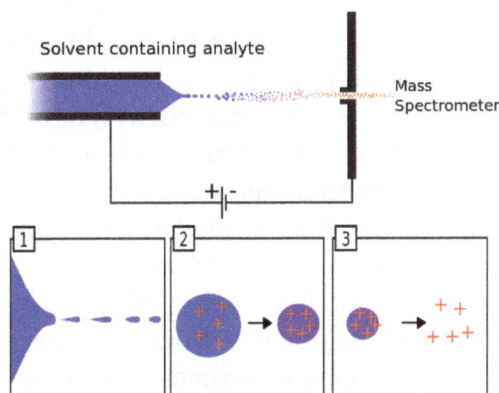

Diagram of Electrospray Ionization. (1) Under high voltage, the Taylor Cone emits a jet of liquid drops (2) The solvent from the droplets progressively evaporates, leaving them more and more charged (3) When the charge exceeds the Rayleigh limit the droplet explosively dissociates, leaving a stream of charged ions

In 1882, Lord Rayleigh theoretically estimated the maximum amount of charge a liquid droplet could carry before throwing out fine jets of liquid. This is now known as the Rayleigh limit.

In 1914, John Zeleny published work on the behaviour of fluid droplets at the end of glass capillaries and presented evidence for different electrospray modes. Wilson and Taylor and Nolan investigated electrospray in the 1920s and Macky in 1931. The electrospray cone (now known as the Taylor cone) was described by Sir Geoffrey Ingram Taylor.

The first use of electrospray ionization with mass spectrometry was reported by Malcolm Dole in 1968.John Bennett Fenn was awarded the 2002 Nobel Prize in Chemistry for the development of electrospray ionization mass spectrometry in the late 1980s.

Ionization Mechanism

Fenn's first electrospray ionization source
coupled to a single quadrupole mass spectrometer

The liquid containing the analyte(s) of interest is dispersed by electrospray, into a fine aerosol. Because the ion formation involves extensive solvent evaporation (also termed desolvation), the typical solvents for electrospray ionization are prepared by mixing water with volatile organic compounds (e.g. methanol acetonitrile). To decrease the initial droplet size, compounds that increase the conductivity (e.g. acetic acid) are customarily added to the solution. These species also act to provide a source of protons to facilitate the ionization process. Large-flow electrosprays can benefit from nebulization a heated inert gas such as nitrogen or carbon dioxide in addition to the high temperature of the ESI source. The aerosol is sampled into the first vacuum stage of a mass spectrometer through a capillary carrying a potential difference of approximately 3000V, which can be heated to aid further solvent evaporation from the charged droplets. The solvent evaporates from a charged droplet until it becomes unstable upon reaching its Rayleigh limit. At this point, the droplet deforms as the electrostatic repulsion of like charges, in an ever-decreasing droplet size, becomes more powerful than the surface tension holding the droplet together. At this point the droplet undergoes Coulomb fission, whereby the original droplet 'explodes' creating many smaller, more stable droplets. The new droplets undergo desolvation and subsequently further Coulomb fissions. During the fission, the droplet loses a small percentage of its mass (1.0–2.3%) along with a relatively large percentage of its charge (10–18%).

There are two major theories that explain the final production of gas-phase ions: the ion evaporation model (IEM) and the charge residue model (CRM). The IEM suggests that as the droplet reaches a certain radius the field strength at the surface of the droplet becomes large enough to assist the field desorption of solvated ions. The CRM suggests that electrospray droplets undergo evaporation and fission cycles, eventually leading progeny droplets that contain on average one analyte ion or less. The gas-phase ions form after the remaining solvent molecules evaporate, leaving the analyte with the charges that the droplet carried.

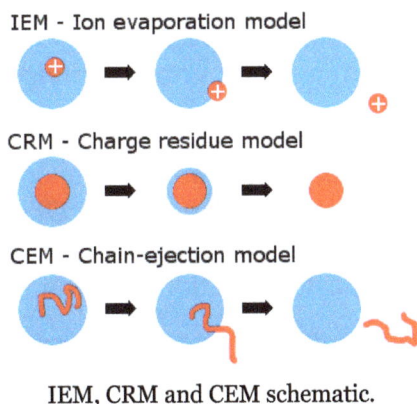

IEM - Ion evaporation model

CRM - Charge residue model

CEM - Chain-ejection model

IEM, CRM and CEM schematic.

A large body of evidence shows either directly or indirectly that small ions (from small molecules) are liberated into the gas phase through the ion evaporation mechanism, while larger ions (from folded proteins for instance) form by charged residue mechanism.

A third model invoking combined charged residue-field emission has been proposed. Another model called chain ejection model (CEM) is proposed for disordered polymers (unfolded proteins).

The ions observed by mass spectrometry may be quasimolecular ions created by the addition of a hydrogen cation and denoted $[M + H]^+$, or of another cation such as sodium ion, $[M + Na]^+$, or the removal of a hydrogen nucleus, $[M - H]^-$. Multiply charged ions such as $[M + nH]^{n+}$ are often observed. For large macromolecules, there can be many charge states, resulting in a characteristic charge state envelope. All these are even-electron ion species: electrons (alone) are not added or removed, unlike in some other ionization sources. The analytes are sometimes involved in electrochemical processes, leading to shifts of the corresponding peaks in the mass spectrum. This effect is demonstrated in the direct ionization of noble metals such as copper, silver and gold using electrospray.

Variants

The electrosprays operated at low flow rates generate much smaller initial droplets, which ensure improved ionization efficiency. In 1993 Gale and Richard D. Smith reported significant sensitivity increases could be achieved using lower flow rates, and down to 200 nL/min. In 1994, two research groups coined the name micro-electrospray (microspray) for electrosprays working at low flow rates. Emmett and Caprioli demonstrated improved performance for HPLC-MS analyses when the electrospray was operated at 300–800 nL/min. Wilm and Mann demonstrated that a capillary flow of ~ 25 nL/min can sustain an electrospray at the tip of emitters fabricated by pulling glass capillaries to a few micrometers. The latter was renamed nano-electrospray (nanospray) in 1996. Currently the name nanospray is also in use for electrosprays fed by pumps at low flow rates, not only for self-fed electrosprays. Although there may not be a well-defined flow rate range for electrospray, microspray, and nano-electrospray, studied "changes in analyte partition during droplet fission prior to ion release". In this paper, they compare results obtained by three other groups. and then measure the signal intensity ratio $[Ba^{2+} + Ba^+]/[BaBr^+]$ at different flow rates.

Cold spray ionization is a form of electrospray in which the solution containing the sample is forced through a small cold capillary (10-80 °C) into an electric field to create a fine mist of cold charged droplets. Applications of this method include the analysis of fragile molecules and guest-host interactions that cannot be studied using regular electrospray ionization.

Electrospray ionization has also been achieved at pressures as low as 25 torr and termed subambient pressure ionization with nanoelectrospray (SPIN) based upon a two-stage ion funnel interface developed by Richard D. Smith and coworkers. The SPIN implementation provided increased sensitivity due to the use of ion funnels that helped confine and transfer ions to the lower pressure region of the mass spectrometer. Operation at low pressure was particularly effective for low flow rates where the smaller electrospray droplet size allowed effective desolvation and ion formation to be achieved. As a result, the researchers were later able to demonstrate achieving an excess of 50% overall ionization utilization efficiency for transfer of ions from the liquid phase, into the gas phase as ions, and through the dual ion funnel interface to the mass spectrometer.

Ambient Ionization

In ambient ionization, the formation of ions occurs outside the mass spectrometer without sample preparation. Electrospray is used for ion formation is a number of ambient ion sources.

Desorption electrospray ionization (DESI) is an ambient ionization technique in which a solvent electrospray is directed at a sample. The electrospray is attracted to the surface by applying a voltage to the sample. Sample compounds are extracted into the solvent which is again aerosolized as highly charged droplets that evaporate to form highly charged ions. After ionization, the ions enter the atmospheric pressure interface of the mass spectrometer. DESI allows for ambient ionization of samples at atmospheric pressure, with little sample preparation.

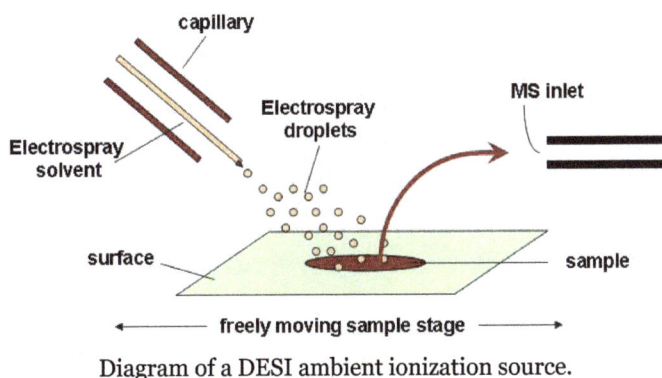

Diagram of a DESI ambient ionization source.

Extractive electrospray ionization is an spray-type, ambient ionization method that uses two merged sprays, one of which is generated by electrospray.

Laser-based electrospray-based ambient ionization is a two-step process in which a pulsed laser is used to desorb or ablate material from a sample and the plume of material interacts with an electrospray to create ions. For ambient ionization, the sample material is deposited on a target near the electrospray. The laser desorbs or ablates material from the sample which is ejected from the surface and into the electrospray which produces highly charged ions. Examples are electrospray laser desorption ionization, matrix-assisted laser desorption electrospray ionization, and laser ablation electrospray ionization

Applications

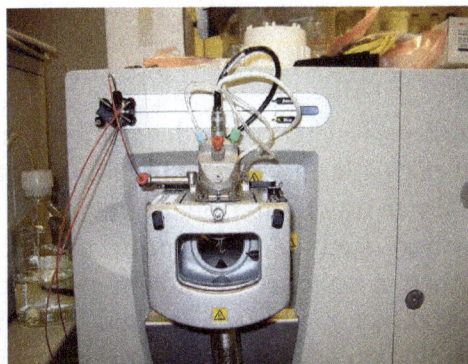

The outside of the electrospray interface on an LTQ mass spectrometer.

Electrospray is used to study protein folding. Additionally, ESI-MS is used to test for the presence of nano clusters, for example U-60.

Liquid Chromatography–mass Spectrometry (LC-MS)

Electrospray ionization is the ion source of choice to couple liquid chromatography with mass spectrometry. The analysis can be performed online, by feeding the liquid eluting from the LC column directly to an electrospray, or offline, by collecting fractions to be later analyzed in a classical nanoelectrospray-mass spectrometry setup. Among the numerous operating parameters in ESI-MS, the electrospray voltage has been identified as an important parameter to consider in ESI LC/MS gradient elution. The effect of various solvent compositions (such as TFA or ammonium acetate, or supercharging reagents, or derivitizing groups) or spraying conditions on Electrospray-LCMS spectra and/or nanoESI-MS spectra.have been studied.

Capillary Electrophoresis-mass Spectrometry (CE-MS)

Capillary electrophoresis-mass spectrometry was enabled by an ESI interface that was developed and patented by Richard D. Smith and coworkers at Pacific Northwest National Laboratory, and shown to have broad utility for the analysis of very small biological and chemical compound mixtures, and even extending to a single biological cell.

Noncovalent Gas Phase Interactions

Electrospray ionization is also utilized in studying noncovalent gas phase interactions. The electrospray process is thought to be capable of transferring liquid-phase noncovalent complexes into the gas phase without disrupting the noncovalent interaction. Problems such as non specific interactions have been identified when studying ligand substrate complexes by ESI-MS or nanoESI-MS. An interesting example of this is studying the interactions between enzymes and drugs which are inhibitors of the enzyme. Competition studies between STAT6 and inhibitors have used ESI as a way to screen for potential new drug candidates.

Matrix-assisted Laser Desorption/ionization

In mass spectrometry, matrix-assisted laser desorption/ionization (MALDI) is an ionization technique that uses a laser energy absorbing matrix to create ions from large molecules with minimal fragmentation. It has been applied to the analysis of biomolecules (biopolymers such as DNA, proteins, peptides and sugars) and large organicmolecules (such as polymers, dendrimers and other macromolecules), which tend to be fragile and fragment when ionized by more conventional ionization methods. It is similar in character to electrospray ionization (ESI) in that both techniques are relatively soft (low fragmentation) ways of obtaining ions of large molecules in the gas phase, though MALDI typically produces far fewer multiply charged ions.

MALDI TOF mass spectrometer

MALDI methodology is a three-step process. First, the sample is mixed with a suitable matrix material and applied to a metal plate. Second, a pulsed laser irradiates the sample, triggering ablation and desorption of the sample and matrix material. Finally, the analyte molecules are ionized by being protonated or deprotonated in the hot plume of ablated gases, and can then be accelerated into whichever mass spectrometer is used to analyse them.

History

The term matrix-assisted laser desorption ionization (MALDI) was coined in 1985 by Franz Hillenkamp, Michael Karas and their colleagues. These researchers found that the amino acidalanine could be ionized more easily if it was mixed with the amino acid tryptophan and irradiated with a pulsed 266 nm laser. The tryptophan was absorbing the laser energy and helping to ionize the non-absorbing alanine. Peptides up to the 2843 Da peptide melittin could be ionized when mixed with this kind of "matrix". The breakthrough for large molecule laser desorption ionization came in 1987 when Koichi Tanaka of Shimadzu Corporation and his co-workers used what they called the "ultra fine metal plus liquid matrix method" that combined 30 nm cobalt particles in glycerol with a 337 nm nitrogen laser for ionization. Using this laser and matrix combination, Tanaka was able to ionize biomolecules as large as the 34,472 Da protein carboxypeptidase-A. Tanaka received one-quarter of the 2002 Nobel Prize in Chemistry for demonstrating that, with the proper combination of laser wavelength and matrix, a protein can be ionized. Karas and Hillenkamp were subsequently able to ionize the 67 kDa protein albumin using a nicotinic acid matrix and a 266 nm laser. Further improvements were realized through the use of a 355 nm laser and the cinnamic acid derivatives ferulic acid, caffeic acid and sinapinic acid as the matrix. The availability of small and relatively inexpensive nitrogen lasers operating at 337 nm wavelength and the first commercial instruments introduced in the early 1990s brought MALDI to an increasing number of researchers. Today, mostly organic matrices are used for MALDI mass spectrometry.

Matrix

UV MALDI Matrix List				
Compound	**Other Names**	**Solvent**	**Wavelength (nm)**	**Applications**
2,5-dihydroxy benzoic acid	DHB, Gentisic acid	acetonitrile, water, methanol, acetone, chloroform	337, 355, 266	peptides, nucleotides, oligonucleotides, oligosaccharides
3,5-dimethoxy-4-hydroxy-cinnamic acid	sinapic acid; sinapinic acid; SA	acetonitrile, water, acetone, chloroform	337, 355, 266	peptides, proteins, lipids
4-hydroxy-3-methoxycinnamic acid	ferulic acid	acetonitrile, water, propanol	337, 355, 266	proteins
α-Cyano-4-hydroxy-cinnamic acid	CHCA	acetonitrile, water, ethanol, acetone	337, 355	peptides, lipids, nucleotides
Picolinic acid	PA	Ethanol	266	oligonucleotides
3-hydroxy picolinic acid	HPA	Ethanol	337, 355	oligonucleotides

The matrix consists of crystallized molecules, of which the three most commonly used are 3,5-dimethoxy-4-hydroxycinnamic acid (sinapinic acid), α-cyano-4-hydroxycinnamic acid (α-CHCA, alpha-cyano or alpha-matrix) and 2,5-dihydroxybenzoic acid (DHB). A solution of one of these molecules is made, often in a mixture of highly purified water and an organic solvent such as acetonitrile (ACN) or ethanol. A counter ion source such as Trifluoroacetic acid (TFA) is usually added to generate the [M+H] ions. A good example of a matrix-solution would be 20 mg/mL sinapinic acid in ACN:water:TFA (50:50:0.1).

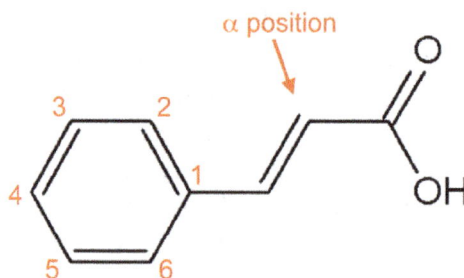

Notation for cinnamic acid substitutions.

The identification of suitable matrix compounds is determined to some extent by trial and error, but they are based on some specific molecular design considerations. They are of a fairly low molecular weight (to allow easy vaporization), but are large enough (with a low enough vapor pressure) not to evaporate during sample preparation or while standing in the mass spectrometer. They are often acidic, therefore act as a proton source to encourage ionization of the analyte. Basic matrices have also been reported. They have a strong optical absorption in either the UV or IR range, so that they rapidly and efficiently absorb the laser irradiation. This efficiency is commonly associated with chemical structures incorporating several conjugated double bonds, as seen in the structure of cinnamic acid. They are functionalized with polar groups, allowing their use in aqueous solutions. They typically contain a chromophore.

The matrix solution is mixed with the analyte (e.g. protein-sample). A mixture of water and organic solvent allows both hydrophobic and water-soluble (hydrophilic) molecules to dissolve into the solution. This solution is spotted onto a MALDI plate (usually a metal plate designed for this purpose). The solvents vaporize, leaving only the recrystallized matrix, but now with analyte molecules embedded into MALDI crystals. The matrix and the analyte are said to be co-crystallized. Co-crystallization is a key issue in selecting a proper matrix to obtain a good quality mass spectrum of the analyte of interest.

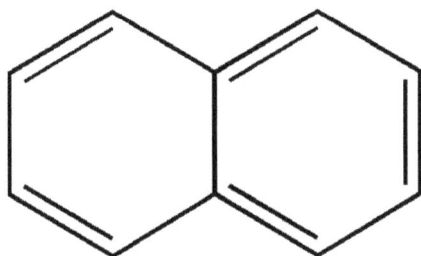

Naphthalene and naphthalene-like compounds can also be used as a matrix to ionize a sample

In analysis of biological systems, Inorganic salts, which are also part of protein extracts, interfere with the ionization process. The salts can be removed by solid phase extraction or by washing the dried-droplet MALDI spots with cold water. Both methods can also remove other substances from the sample. The matrix-protein mixture is not homogenous because the polarity difference leads to a separation of the two substances during co-crystallization. The spot diameter of the target is much larger than that of the laser, which makes it necessary to make many laser shots at different places of the target, to get the statistical average of the substance concentration within the target spot.

The matrix can be used to tune the instrument to ionize the sample in different ways. As mentioned above, acid-base like reactions are often utilized to ionize the sample, however, molecules with conjugated pi systems, such as naphthalene like compounds, can also serve as an electron acceptor and thus a matrix for MALDI/TOF. This is particularly useful in studying molecules that also possess conjugated pi systems.The most widely used application for these matrices is studying porphyrin like compounds such as chlorophyll. These matrices have been shown to have better ionization patterns that do not result in odd fragmentation patterns or complete loss of side chains. It has also been suggested that conjugated porphryin like molecules can serve as a matrix and cleave themselves eliminating the need for a separate matrix compound.

Instrumentation

There are several variations of the MALDI technology and comparable instruments are today produced for very different purposes. From more academic and analytical, to more industrial and high throughput. The MS field has expanded into requiring ultrahigh resolution mass spectrometry such as the FT-ICR instruments as well as more high-throughput instruments. As many MALDI MS instruments can be bought with an interchangeable ionization source (Electrospray ionization, MALDI, Atmospheric pressure ionization, etc.) the technologies often overlap and many times any soft ionization method could potentially be used. For more variations of soft ionization methods go to Soft laser desorption or Ion source.

Laser

MALDI techniques typically employ the use of UV lasers such as nitrogen lasers (337 nm) and frequency-tripled and quadrupled Nd:YAG lasers (355 nm and 266 nm respectively).

Infrared laser wavelengths used for infrared MALDI include the 2.94 μm Er:YAG laser, mid-IR optical parametric oscillator,and 10.6 μm carbon dioxide laser. Although not as common, infrared lasers are used due to their softer mode of ionization. IR-MALDI also has the advantage of greater material removal (useful for biological samples), less low-mass interferences, and compatibility with other matrix-free laser desorption mass spectrometry methods.

Time of Flight

Sample target for a MALDI mass spectrometer

The type of a mass spectrometer most widely used with MALDI is the TOF (time-of-flight mass spectrometer), mainly due to its large mass range. The TOF measurement procedure is also ideally suited to the MALDI ionization process since the pulsed laser takes individual 'shots' rather than working in continuous operation. MALDI-TOF instrument or reflectron is equipped with an "ion mirror" that reflects ions using an electric field, thereby doubling the ion flight path and increasing the resolution. Today, commercial reflectron TOF instruments reach a resolving power $m/\Delta m$ of well above 20,000 FWHM (full-width half-maximum, Δm defined as the peak width at 50% of peak height).

MALDI has been coupled with IMS-TOF MS to identify phosphorylated and non-phosphorylated peptides.

MALDI-FT-ICR MS has been demonstrated to be a useful technique where high resolution MALDI-MS measurements are desired.

Atmospheric Pressure

Atmospheric pressure (AP) matrix-assisted laser desorption/ionization (MALDI) is an ionization technique (ion source) that in contrast to vacuum MALDI operates at normal atmospheric environment. The main difference between vacuum MALDI and AP-MALDI is the pressure in which the ions are created. In vacuum MALDI, ions are typically produced at 10 mTorr or less while in AP-MALDI ions are formed in atmospheric pressure. In the past the main disadvantage of AP MALDI technique compared to the conventional vacuum MALDI has been its limited sensitivity; however, ions can be transferred into the mass spectrometer with high efficiency and attomole detection limits have been reported. AP-MALDI is used in mass spectrometry (MS) in a variety of applications ranging from proteomics to drug discovery. Popular topics that are addressed by AP-MALDI mass spectrometry include: proteomics; mass analysis of DNA, RNA, PNA, lipids, oligosaccharides, phosphopeptides, bacteria, small molecules and synthetic polymers, similar applications as available also for vacuum MALDI instruments. The AP-MALDI ion source is easily coupled to an ion trap mass spectrometer or any other MS system equipped with ESI (electrospray ionization) or nanoESI source.

Ionization Mechanism

The laser is fired at the matrix crystals in the dried-droplet spot. The matrix absorbs the laser energy and it is thought that primarily the matrix is desorbed and ionized (by addition of a proton) by this event. The hot plume produced during ablation contains many species: neutral and ionized matrix molecules, protonated and deprotonated matrix molecules, matrix clusters and nanodroplets. Ablated species may participate in the ionization of analyte, though the mechanism of MALDI is still debated. The matrix is then thought to transfer protons to the analyte molecules (e.g., protein molecules), thus charging the analyte. An ion observed after this process will consist of the initial neutral molecule [M] with ions added or removed. This is called a quasimolecular ion, for example $[M+H]^+$ in the case of an added proton, $[M+Na]^+$ in the case of an added sodium ion, or $[M-H]^-$ in the case of a removed proton. MALDI is capable of creating singly charged ions or multiply charged ions ($[M+nH]^{n+}$) depending on the nature of the matrix, the laser intensity, and/or the voltage used. Note that these are all even-electron species. Ion signals of radical cations (photoionized molecules) can be observed, e.g., in the case of matrix molecules and other organic molecules.

The gas phase proton transfer model, implemented as the coupled physical and chemical dynamics (CPCD) model, of UV laser MALDI postulates primary and secondary processes leading to ionization. Primary processes involve initial charge separation through absorption of photons by the matrix and pooling of the energy to form matrix ion pairs. Primary ion formation occurs through absorption of a UV photon to create excited state molecules by

$$S_0 + h\nu \rightarrow S_1$$

$$S_1 + S_1 \rightarrow S_0 + S_n$$

$$S_1 + S_n \rightarrow M^+ + M^-$$

where S_0 is the ground electronic state, S_1 the first electronic excited state, and S_n is a higher electronic excited state. The product ions can be proton transfer or electron transfer ion pairs, indicated by M^+ and M^- above. Secondary processes involve ion-molecule reactions to form analyze ions.

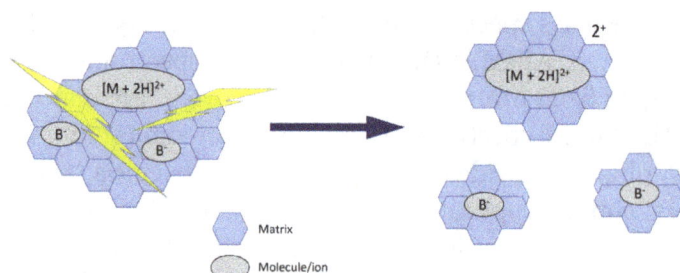

In the lucky survivor model, positive ions can be formed from highly charged clusters produced during break-up of the matrix- and analyte-containing solid.

The lucky survivor model (cluster ionization mechanism) postulates that analyze molecules are incorporated in the matrix maintaining the charge state from solution. Ion formation occurs through charge separation upon fragmentation of laser ablated clusters. Ions that are not neutralized by recombination with photoelectrons or counter ions are the so called lucky survivors.

The thermal model postulates that the high temperature facilitates the proton transfer between matrix and analyte in melted matrix liquid. Ion-to-neutral ratio is an important parameter to justify the theoretical model, and the mistaken citation of ion-to-neutral ratio could result in an erroneous determination of the ionization mechanism. The model quantitatively predicts the increase in total ion intensity as a function of the concentration and proton affinity of the analytes, and the ion-to-neutral ratio as a function of the laser fluences. This model also suggests that metal ion adducts (e.g., $[M+Na]^+$ or $[M+K]^+$) are mainly generated from the thermally induced dissolution of salt.

The matrix-assisted ionization (MAI) method uses matrix preparation similar to MALDI but does not require laser ablation to produce analyte ions of volatile or non-volatile compounds. Simply exposing the matrix with analyte to the vacuum of the mass spectrometer creates ions with nearly identical charge states to electrospray ionization. It is suggested that there are likely mechanistic commonality between this process and MALDI.

Applications

Biochemistry

In proteomics, MALDI is used for the rapid identification of proteins isolated by using gel electrophoresis: SDS-PAGE, size exclusion chromatography, affinity chromatography, strong/weak ion exchange, isotope coded protein labelling (ICPL),and two-dimensional gel electrophoresis. Peptide mass fingerprinting is the most popular analytical application of MALDI-TOF mass spectrometers. MALDI TOF/TOF mass spectrometers are used to reveal amino acid sequence of peptides using post-source decay or high energy collision-induced dissociation.

Loss of sialic acid has been identified in papers when DHB has been used as a matrix for MALDI MS analysis of glycosylated peptides. Using sinapinic acid, 4-HCCA and DHB as matrices, S. Martin studied loss of sialic acid in glycosylated peptides by metastable decay in MALDI/TOF in linear mode and reflector mode. A group at Shimadzu Corporation derivatized the sialic acid by an amidation reaction as a way to improve detection sensitivity and also demonstrated that ionic liquid matrix reduces a loss of sialic acid during MALDI/TOF MS analysis of sialylated oligosaccharides. THAP, DHAP, and a mixture of 2-aza-2-thiothymine and phenylhydrazine have been identified as matrices that could be used to minimize loss of sialic acid during MALDI MS analysis of glycosylated peptides.

It has been reported that a reduction in loss of some post-translational modifications can be accomplished if IR MALDI is used instead of UV MALDI.

In molecular biology, a mixture of 5-methoxysalicylic acid and spermine can be used as a matrix for oligonucleotides analysis in MALDI mass spectrometry, for instance after oligonucleotide synthesis.

Organic Chemistry

Some synthetic macromolecules, such as catenanes and rotaxanes, dendrimers and hyperbranched polymers, and other assemblies, have molecular weights extending into the thousands or tens of thousands, where most ionization techniques have difficulty producing molecular ions. MALDI is a simple and fast analytical method that can allow chemists to rapidly analyze the results of such syntheses and verify their results.

Polymers

In polymer chemistry MALDI can be used to determine the molar mass distribution. Polymers with polydispersity greater than 1.2 are difficult to characterize with MALDI due to the signal intensity discrimination against higher mass oligomers.

A good matrix for polymers is dithranol or AgTFA. The sample must first be mixed with dithranol and the AgTFA added afterwards; otherwise the sample would precipitate out of solution.

Microbiology

MALDI/TOF spectra are used for the identification of micro-organisms such as bacteria or fungi. A portion of a colony of the microbe in question is placed onto the sample target and overlaid with matrix. The mass spectra generated are analyzed by dedicated software and compared with stored profiles. Species diagnosis by this procedure is much faster, more accurate and cheaper than other procedures based on immunological or biochemical tests. MALDI/TOF is becoming a standard method for species identification in medical microbiological laboratories.

Its main advantage over other microbiological identification methods is its ability to reliably identify, at low cost and rapidly, a wide variety of micro-organisms directly from the selective medium used to isolate/detect them. The absence of the need to purify the suspect (or "presumptive") colony allowing for much faster turn-around times.

Medicine

MALDI/TOF spectra are often utilized in tandem with other analysis and spectroscopy techniques in the diagnosis of diseases. MALDI/TOF is a diagnostic tool with much potential because it allows for the rapid identification of proteins and changes to proteins without the cost or computing power of sequencing nor the skill or time needed to solve a crystal structure in X-ray crystallography.

One of example of this is necrotizing enterocolitis (NEC), which is a devastating disease that affects the bowels of premature infants. The symptoms of NEC are very similar to those of sepsis, and many infants die awaiting diagnosis and treatment. MALDI/TOF was used to quickly analyze fecal samples and find differences between the mutant and the functional protein responsible for NEC. There is hope that a similar technique could be used as a quick, diagnostic tool that would not require sequencing.

Another example of the diagnostic power of MALDI/TOF is in the area of cancer. Pancreatic cancer remains one of the most deadly and difficult to diagnose cancers. Impaired cellular signaling due to mutations in membrane proteins has been long suspected to contribute to pancreatic cancer. MALDI/TOF has been used to identify a membrane protein associated with pancreatic cancer and at one point may even serve as an early detection technique.

MALDI/TOF can also potentially be used to dictate treatment as well as diagnosis. MALDI/TOF serves as a method for determining the drug resistance of bacteria, especially to β-lactamases (Penicillin family). The MALDI/TOF detects the presence of carbapenemases, which indicates drug resistance to standard antibiotics. It is predicted that this could serve as a method for identifying a bacterium as drug resistant in as little as three hours. This technique could help physicians decide whether to prescribe more aggressive antibiotics initially.

References

- Lu, L; Horstmann, H; Ng, C; Hong, W (December 2001). "Regulation of Golgi structure and function by ARF-like protein 1 (Arl1).". Journal of Cell Science. 114 (Pt 24): 4543–55. PMID 11792819

- Ninfa, Alexander J.; Ballou, David P.; Benore, Marilee (2009). Fundamental Laboratory Approaches for Biochemistry and Biotechnology (2 ed.). Wiley. p. 133. ISBN 9780470087664

- Karas, Michael; Krüger, Ralf (2003). "Ion Formation in MALDI: The Cluster Ionization Mechanism". Chemical Reviews. 103 (2): 427–440. ISSN 0009-2665. doi:10.1021/cr010376a

- Bouveret E, et al. (2000). "A Sm-like protein complex that participates in mRNA degradation". The EMBO Journal. 19 (7): 1661–1671. PMC 310234. PMID 10747033. doi:10.1093/emboj/19.7.1661

- M., Grisham, Charles (2013-01-01). Biochemistry. Brooks/Cole, Cengage Learning. ISBN 1133106293. OCLC 777722371

- Markides, K.; Gräslund, A. (9 October 2002). "Advanced information on the Nobel Prize in Chemistry 2002" (PDF). The Royal Swedish Academy of Sciences. pp. 1–13. Retrieved 2013-08-28

- Dole M, Mack LL, Hines RL, Mobley RC, Ferguson LD, Alice MB (1968). "Molecular Beams of Macroions". Journal of Chemical Physics. 49 (5): 2240–2249. Bibcode:1968JChPh..49.2240D. doi:10.1063/1.1670391

- Birendra N. Pramanik; A.K. Ganguly; Michael L. Gross (28 February 2002). Applied Electrospray Mass Spectrometry: Practical Spectroscopy Series. CRC Press. pp. 4–. ISBN 978-0-8247-4419-9

- Laiko, V. V.; Moyer, S. C.; Cotter, R. J. (2000). "Atmospheric pressure MALDI/ion trap mass spectrometry". Analytical Chemistry. 72 (21): 5239–43. PMID 11080870. doi:10.1021/ac000530d

- Dole M, Mack LL, Hines RL, Mobley RC, Ferguson LD, Alice MB (1968). "Molecular Beams of Macroions". Journal of Chemical Physics. 49 (5): 2240–2249. Bibcode:1968JChPh..49.2240D. doi:10.1063/1.1670391

- Strupat, K.; Scheibner, O.; Arrey, T.; Bromirski, M. (2011). "Biological Applications of AP MALDI with Thermo Scientific Exactive Orbitrap MS" (PDF). 443-539-1710. Thermo Scientific. Retrieved 17 June 2011

- Birendra N. Pramanik; A.K. Ganguly; Michael L. Gross (28 February 2002). Applied Electrospray Mass Spectrometry: Practical Spectroscopy Series. CRC Press. pp. 4–. ISBN 978-0-8247-4419-9

- Harvey, D. J. (1999). "Matrix-assisted laser desorption/ionization mass spectrometry of carbohydrates". Mass Spectrometry Reviews. 18 (6): 349–451. doi:10.1002/(SICI)1098-2787(1999)18:6<349::AID-MAS1>3.0.CO;2-H

- Korfmacher, Walter A. (2009). Using Mass Spectrometry for Drug Metabolism Studies. CRC Press. p. 342. ISBN 9781420092219

3

High-throughput Techniques:
A Comprehensive Study

High-throughput method is required when studying large number of proteins simultaneously. Protein microarray is a method to study proteins. Antibody microarrays have also been discussed here. This chapter provides a plethora of interdisciplinary topics for better comprehension of protein-protein interactions.

Protein Microarrays

Although the traditional methods of studying protein-protein interactions are effective, they are very time consuming and thus, limit the number of proteins that can be studied at a time. For studying large number of proteins simultaneously, high-throughput methods are needed. One of the high-throughput methods developed for studying protein-protein interaction is protein microarrays. It is one of the most convenient methods for studying large number of proteins in a single experiment. Protein microarrays have a solid surface like a chip or a glass slide, on which thousands of proteins are spotted. The array is incubated with specific probe molecules, which are tagged with a reporter molecule such as a fluorescent chromophore. The array is scanned using a laser or any appropriate method of detecting the reporter and the interaction can be identified as the probes bind at the appropriate proteins spots. In this way, thousands of targets can be screened in one go. This high-throughput platform can be used for biomarker discovery, antigen-antibody studies, studying protein-protein interactions, identification of new interactions and for functional characterization of interacting partners.

A protein microarray (or protein chip) is a high-throughput method used to track the interactions and activities of proteins, and to determine their function, and determining function on a large scale. Its main advantage lies in the fact that large numbers of proteins can be tracked in parallel. The chip consists of a support surface such as a glass slide, nitrocellulose membrane, bead, or microtitre plate, to which an array of capture proteins is bound. Probe molecules, typically labeled with a fluorescent dye, are added to the array. Any reaction between the probe and the immobilised protein emits a fluorescent signal that is read by a laser scanner. Protein microarrays are rapid, automated, economical, and highly sensitive, consuming small quantities of samples and reagents. The concept and methodology of protein microarrays was first introduced and illustrated in antibody microarrays (also referred to as antibody matrix) in 1983 in a scientific publication and a series of patents. The high-throughput technology behind the protein microarray was relatively easy to develop since it is based on the technology developed for DNA microarrays, which have become the most widely used microarrays.

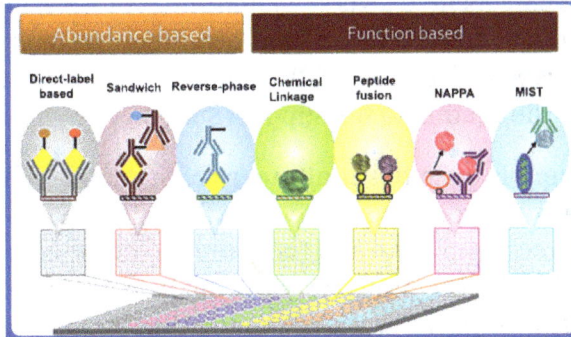

Various types of Protein Microarrays: The protein microarrays can be broadly classified into two kinds, abundance-based and function-based protein microarrays. The abundance-based protein microarrays consist of direct label-based, sandwich and reverse-phase protein microarrays. While the function-based protein microarrays consist of peptide fusion, NAPPA (nucleic acid programmable protein array) and MIST (multiple spotting technique).

Motivation for Development

Protein microarrays were developed due to the limitations of using DNA microarrays for determining gene expression levels in proteomics. The quantity of mRNA in the cell often doesn't reflect the expression levels of the proteins they correspond to. Since it is usually the protein, rather than the mRNA, that has the functional role in cell response, a novel approach was needed. Additionally post-translational modifications, which are often critical for determining protein function, are not visible on DNA microarrays. Protein microarrays replace traditional proteomics techniques such as 2D gel electrophoresis or chromatography, which were time consuming, labor-intensive and ill-suited for the analysis of low abundant proteins.

Making the Array

The proteins are arrayed onto a solid surface such as microscope slides, membranes, beads or microtitre plates. The function of this surface is to provide a support onto which proteins can be immobilized. It should demonstrate maximal binding properties, whilst maintaining the protein in its native conformation so that its binding ability is retained. Microscope slides made of glass or silicon are a popular choice since they are compatible with the easily obtained robotic arrayers and laser scanners that have been developed for DNA microarray technology. Nitrocellulose film slides are broadly accepted as the highest protein binding substrate for protein microarray applications.

The chosen solid surface is then covered with a coating that must serve the simultaneous functions of immobilising the protein, preventing its denaturation, orienting it in the appropriate direction so that its binding sites are accessible, and providing a hydrophilic environment in which the binding reaction can occur. In addition, it also needs to display minimal non-specific binding in order to minimize background noise in the detection systems. Furthermore, it needs to be compatible with different detection systems. Immobilising agents include layers of aluminium or gold, hydrophilic polymers, and polyacrylamide gels, or treatment with amines, aldehyde or epoxy. Thin-film technologies like physical vapour deposition (PVD) and chemical vapour deposition (CVD) are employed to apply the coating to the support surface.

An aqueous environment is essential at all stages of array manufacture and operation to prevent protein denaturation. Therefore, sample buffers contain a high percent of glycerol(to lower the freezing point), and the humidity of the manufacturing environment is carefully regulated. Microwells have the dual advantage of providing an aqueous environment while preventing cross-contamination between samples.

In the most common type of protein array, robots place large numbers of proteins or their ligands onto a coated solid support in a pre-defined pattern. This is known as robotic contact printing or robotic spotting. Another fabrication method is ink-jetting, a drop-on-demand, non-contact method of dispersing the protein polymers onto the solid surface in the desired pattern.Piezoelectric spotting is a similar method to ink-jet printing. The printhead moves across the array, and at each spot uses electric stimulation to deliver the protein molecules onto the surface via tiny jets. This is also a non-contact process.Photolithography is a fourth method of arraying the proteins onto the surface. Light is used in association with photomasks, opaque plates with holes or transparencies that allow light to shine through in a defined pattern. A series of chemical treatments then enables deposition of the protein in the desired pattern upon the material underneath the photomask.

The capture molecules arrayed on the solid surface may be antibodies, antigens, aptamers (nucleic acid-based ligands), affibodies (small molecules engineered to mimic monoclonal antibodies), or full length proteins. Sources of such proteins include cell-based expression systems for recombinant proteins, purification from natural sources, production in vitro by cell-free translation systems, and synthetic methods for peptides. Many of these methods can be automated for high throughput production but care must be taken to avoid conditions of synthesis or extraction that result in a denatured protein which, since it no longer recognizes its binding partner, renders the array useless.

Proteins are highly sensitive to changes in their microenvironment. This presents a challenge in maintaining protein arrays in a stable condition over extended periods of time. In situ methods involve on-chip synthesis of proteins as and when required, directly from the DNA using cell-free protein expression systems. Since DNA is a highly stable molecule it does not deteriorate over time and is therefore suited to long-term storage. This approach is also advantageous in that it circumvents the laborious and often costly processes of separate protein purification and DNA cloning, since proteins are made and immobilised simultaneously in a single step on the chip surface. Examples of in situ techniques are PISA (protein in situ array), NAPPA (nucleic acid programmable protein array) and DAPA (DNA array to protein array).

Types of Arrays

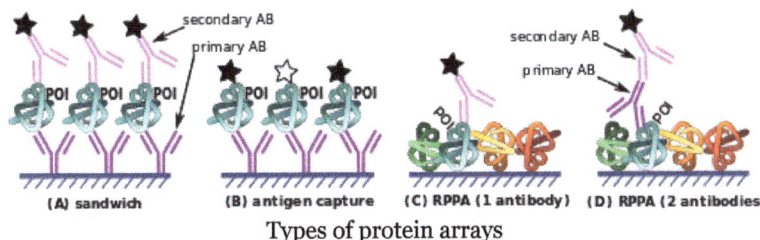

Types of protein arrays

There are three types of protein microarrays that are currently used to study the biochemical activities of proteins.

Analytical microarrays are also known as capture arrays. In this technique, a library of antibodies, aptamers or affibodies is arrayed on the support surface. These are used as capture molecules since each binds specifically to a particular protein. The array is probed with a complex protein solution such as a cell lysate. Analysis of the resulting binding reactions using various detection systems can provide information about expression levels of particular proteins in the sample as well as measurements of binding affinities and specificities. This type of microarray is especially useful in comparing protein expression in different solutions. For instance the response of the cells to a particular factor can be identified by comparing the lysates of cells treated with specific substances or grown under certain conditions with the lysates of control cells. Another application is in the identification and profiling of diseased tissues.

Functional protein microarrays (also known as target protein arrays) are constructed by immobilising large numbers of purified proteins and are used to identify protein–protein, protein–DNA, protein–RNA, protein–phospholipid, and protein–small-molecule interactions, to assay enzymatic activity and to detect antibodies and demonstrate their specificity. They differ from analytical arrays in that functional protein arrays are composed of arrays containing full-length functional proteins or protein domains. These protein chips are used to study the biochemical activities of the entire proteome in a single experiment.

Reverse phase protein microarray (RPPA) involve complex samples, such as tissue lysates. Cells are isolated from various tissues of interest and are lysed. The lysate is arrayed onto the microarray and probed with antibodies against the target protein of interest. These antibodies are typically detected with chemiluminescent, fluorescent or colorimetric assays. Reference peptides are printed on the slides to allow for protein quantification of the sample lysates. RPAs allow for the determination of the presence of altered proteins or other agents that may be the result of disease.

Specifically, post-translational modifications, which are typically altered as a result of disease can be detected using RPAs.

Detection

Protein array detection methods must give a high signal and a low background. The most common and widely used method for detection is fluorescence labeling which is highly sensitive, safe and compatible with readily available microarray laser scanners. Other labels can be used, such as affinity, photochemical or radioisotope tags. These labels are attached to the probe itself and can interfere with the probe-target protein reaction. Therefore, a number of label free detection methods are available, such as surface plasmon resonance (SPR), carbon nanotubes, carbon nanowire sensors (where detection occurs via changes in conductance) and microelectromechanical system (MEMS) cantilevers. All these label free detection methods are relatively new and are not yet suitable for high-throughput protein interaction detection; however, they do offer much promise for the future.

Protein quantitation on nitrocellulose coated glass slides can use near-IR fluorescent detection. This limits interferences due to auto-fluorescence of the nitrocellulose at the UV wavelengths used for standard fluorescent detection probes.

Applications

There are five major areas where protein arrays are being applied: diagnostics, proteomics, protein functional analysis, antibody characterization, and treatment development.

Diagnostics involves the detection of antigens and antibodies in blood samples; the profiling of sera to discover new disease biomarkers; the monitoring of disease states and responses to therapy in personalized medicine; the monitoring of environment and food. Digital bioassay is a en example of using protein microarray for diagnostic purposes. In this technology, an array of microwells on a glass/polymer chip are seeded with magnetic beads (coated with fluorescent tagged antibodies), subjected to targeted antigens and then characterised by a microscope through counting fluorescing wells. A cost-effective fabrication platform (using OSTE polymers) for such microwell arrays has been recently demonstrated and the bio-assay model system has been successfully characterised.

Proteomics pertains to protein expression profiling i.e. which proteins are expressed in the lysate of a particular cell.

Protein functional analysis is the identification of protein–protein interactions (e.g. identification of members of a protein complex), protein–phospholipid interactions, small molecule targets, enzymatic substrates (particularly the substrates of kinases) and receptor ligands.

Antibody characterization is characterizing cross-reactivity, specificity and mapping epitopes.

Treatment development involves the development of antigen-specific therapies for autoimmunity, cancer and allergies; the identification of small molecule targets that could potentially be used as new drugs.

Challenges

Despite the considerable investments made by several companies, proteins chips have yet to flood the market. Manufacturers have found that proteins are actually quite difficult to handle. A protein chip requires a lot more steps in its creation than does a DNA chip.

Challenges include: 1) finding a surface and a method of attachment that allows the proteins to maintain their secondary or tertiary structure and thus their biological activity and their interactions with other molecules, 2) producing an array with a long shelf life so that the proteins on the chip do not denature over a short time, 3) identifying and isolating antibodies or other capture molecules against every protein in the human genome, 4) quantifying the levels of bound protein while assuring sensitivity and avoiding background noise, 5) extracting the detected protein from the chip in order to further analyze it, 6) reducing non-specific binding by the capture agents, 7) the capacity of the chip must be sufficient to allow as complete a representation of the proteome to be visualized as possible; abundant proteins overwhelm the detection of less abundant proteins such as signaling molecules and receptors, which are generally of more therapeutic interest.

Antibody Microarray

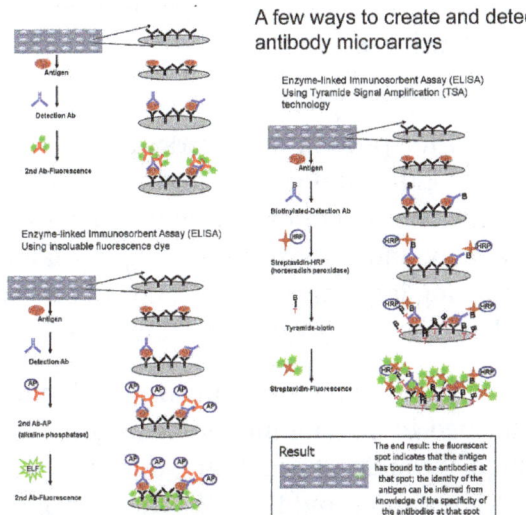

Samples of antibody microarray creations and detections.

An antibody microarray (also known as antibody array) is a specific form of protein microarray. In this technology, a collection of capture antibodies are spotted and fixed on a solid surface such as glass, plastic, membrane, or silicon chip, and the interaction between the antibody and its target antigen is detected. Antibody microarrays are often used for detecting protein expression from various biofluids including serum, plasma and cell or tissue lysates. Antibody arrays may be used for both basic research and medical and diagnostic applications.

Background

The concept and methodology of antibody microarrays were first introduced by Tse Wen Chang in 1983 in a scientific publication and a series of patents, when he was working at Centocor in Malvern, Pennsylvania. Chang coined the term "antibody matrix" and discussed "array" arrangement of minute antibody spots on small glass or plastic surfaces. He demonstrated that a 10×10 (100 in total) and 20×20 (400 in total) grid of antibody spots could be placed on a 1×1 cm surface. He also estimated that if an antibody is coated at a 10 µg/mL concentration, which is optimal for most antibodies, 1 mg of antibody can make 2,000,000 dots of 0.25 mm diameter. Chang's invention focused on the employment of antibody microarrays for the detection and quantification of cells bearing certain surface antigens, such as CD antigens and HLA allotypic antigens, particulate antigens, such as viruses and bacteria, and soluble antigens. The principle of "one sample application, multiple determinations", assay configuration, and mechanics for placing absorbent dots described in the paper and patents should be generally applicable to different kinds of microarrays. When Tse Wen Chang and Nancy T. Chang were setting up Tanox, Inc. in Houston, Texas in 1986, they purchased the rights on the antibody matrix patents from Centocor as part of the technology base to build their new startup. Their first product in development was an assay, termed "immunosorbent cytometry", which could be employed to monitor the immune status, i.e., the concentrations and ratios of $CD3^+$, $CD4^+$, and $CD8^{+T\,cells}$, in the blood of HIV-infected individuals.

The theoretical background for protein microarray-based ligand binding assays was further developed by Roger Ekins and colleagues in the late 1980s. According to the model, antibody microarrays would not only permit simultaneous screening of an analyte panel, but would also be more sensitive and rapid than conventional screening methods. Interest in screening large protein sets only arose as a result of the achievements in genomics by DNA microarrays and the Human Genome Project.

The first array approaches attempted to miniaturize biochemical and immunobiological assays usually performed in 96-well microtiter plates. While 96-well plate-based antibody arrays have high-throughput capability, the small surface area in each well limits the number of antibody spots and thus, the number of analytes detected. Other solid supports, such as glass slides and nitrocellulose membranes, were subsequently utilized to develop arrays which could accommodate larger panels of antibodies. Nitrocellulose membrane-based arrays are flexible, easy to handle, and have increased protein binding capacity, but are less amenable to high throughput or automated processing. Chemically derivatized glass slides allow for printing of sub-microliter sized antibody spots, reducing the array surface area without sacrificing spot density. This in turn reduces the volume of sample consumed. Glass slide-based arrays, owing to their smooth and rigid structure, can also be easily fitted to high-throughput liquid handling systems.

Most antibody array systems employ 1 of 2 non-competitive methods of immunodetection: single-antibody (label-based) detection and 2-antibody (sandwich-based) detection. The latter method, in which analyte detection requires the binding of 2 distinct antibodies (a capture antibody and a reporter antibody, each binding to a unique epitope), confers greater specificity and lower background signal compared with label-based immunodetection (where only 1 capture antibody is used and detection is achieved by chemically labeling all proteins in the starting sample). Sandwich-based antibody arrays usually attain the highest specificity and sensitivity (ng – pg levels) of any array format; their reproducibility also enables quantitative analysis to be performed. Due to the difficulty of developing matched antibody pairs that are compatible with all other antibodies in the panel, small arrays often make use of a sandwich approach. Conversely, high-density arrays are easier to develop at a lower cost using the single antibody label-based approach. In this methodology, one set of specific antibodies is used and all the proteins in a sample are labelled directly by fluorescent dyes or haptens.

Initial uses of antibody-based array systems included detecting IgGs and specific subclasses, analyzing antigens, screening recombinant antibodies,studying yeast protein kinases, analyzing autoimmune antibodies, and examining protein-protein interactions. The first approach to simultaneously detect multiple cytokines from physiological samples using antibody array technology was by Ruo-Pan Huang and colleagues in 2001. Their approach used Hybond ECL membranes to detect a small panel of 24 cytokines from cell culture conditioned media and patient's sera and was able to profile cytokine expression at physiological levels. Huang took this technology and started a new business, RayBiotech, Inc., the first to successfully commercialize a planar antibody array.

In the last ten years, the sensitivity of the method was improved by an optimization of the surface chemistry as well as dedicated protocols for their chemical labeling. Currently, the sensitivity of antibody arrays is comparable to that of ELISA and antibody arrays are regularly used for profiling experiments on tissue samples, plasma or serum samples and many other sample types. One main focus in antibody array based profiling studies is biomarker discovery, specifically for cancer. For cancer-related research, the development and application of an antibody array comprising 810 different cancer-related antibodies was reported in 2010. Also in 2010, an antibody array comprising 507 cytokines, chemokines, adipokines, growth factors, angiogenic factors, proteases, soluble receptors, soluble adhesion molecules, and other proteins was used to screen the serum of ovarian cancer patients and healthy individuals and found a significant difference in protein expression between normal and cancer samples. More recently, antibody arrays have helped determine specific allergy-related serum proteins whose levels are associated with glioma and can reduce the risk years before diagnosis. Protein profiling with antibody arrays have also proven successful in areas other than cancer research, specifically in neurological diseases such as Alzheimer's. A number of studies have attempted to identify biomarker panels that can distinguish Alzheimer's patients, and many have used antibody arrays in this process. Jaeger and colleagues measured nearly 600 circulatory proteins to discover biological pathways and networks affected in Alzheimer's and explored the positive and negative relationships of the levels of those individual proteins and networks with the cognitive performance of Alzheimer's patients. Currently the largest commercially available sandwich-based antibody array detects 1000 different proteins. Antibody arrays are often used for detecting protein expression from many sample types, but also in those with various preparations. Jiang and colleagues illustrated nicely the correlation between array protein expression in two different blood preparations: serum and dried blood spots. These different blood sample preparations were analyzed using three antibody array platforms: sandwich-based, quantitative, and label-based, and a strong correlation in protein expression was found, suggesting that dried blood spots, which are a more convenient, safe, and inexpensive means of obtaining blood especially in non-hospitalized public health areas, can be used effectively with antibody array analysis for biomarker discovery, protein profiling, and disease screening, diagnosis, and treatment.

Applications

Using antibody microarray in different medical diagnostic areas has attracted researchers attention. Digital bioassay is a en example of such research domains. In this technology, an array of microwells on a glass/polymer chip are seeded with magnetic beads (coated with fluorescent tagged antibodies), subjected to targeted antigens and then characterised by a microscope through counting fluorescing wells. A cost-effective fabrication platform (using OSTE polymers) for such microwell arrays has been recently demonstrated and the bio-assay model system has been successfully characterised.

Reverse Phase Protein Lysate Microarray

Reverse phase protein array (RPPA) is a protein array designed as a micro- or nano-scaled dot-blot platform that allows measurement of protein expression levels in a large number of biological samples simultaneously in a quantitative manner when high-quality antibodies are available.

Technically, minuscule amounts of a) cellular lysates, from intact cells or laser capture microdissected cells, b) body fluids such as serum, CSF, urine, vitreous, saliva, etc., are immobilized on individual spots on a microarray that is then incubated with a single specific antibody to detect expression of the target protein across many samples. One microarray, depending on the design, can accommodate hundreds to thousands of samples that are printed in a series of replicates. Detection is performed using either a primary or a secondary labeled antibody by chemiluminescent, fluorescent or colorimetric assays. The array is then imaged and the obtained data is quantified.

Multiplexing is achieved by probing multiple arrays spotted with the same lysate with different antibodies simultaneously and can be implemented as a quantitative calibrated assay. In addition, since RPMA can utilize whole-cell or undissected or microdissected cell lysates, it can provide direct quantifiable information concerning post translationally modified proteins that are not accessible with other high-throughput techniques. Thus, RPMA provides high-dimensional proteomic data in a high throughput, sensitive and quantitative manner. However, since the signal generated by RPMA could be generated from unspecific primary or secondary antibody binding, as is seen in other techniques such as ELISA, or immunohistochemistry, the signal from a single spot could be due to cross-reactivity. Thus, the antibodies used in RPMA must be carefully validated for specificity and performance against cell lysates by western blot.

RPMA has various uses such as quantitative analysis of protein expression in cancer cells, body fluids or tissues for biomarker profiling, cell signaling analysis and clinical prognosis, diagnosis or therapeutic prediction. This is possible as a RPMA with lysates from different cell lines and or laser capture microdissected tissue biopsies of different disease stages from various organs of one or many patients can be constructed for determination of relative or absolute abundance or differential expression of a protein marker level in a single experiment. It is also used for monitoring protein dynamics in response to various stimuli or doses of drugs at multiple time points. Some other applications that RPMA is used for include exploring and mapping protein signaling pathways, evaluating molecular drug targets and understanding a candidate drug's mechanism of action. It has been also suggested as a potential early screen test in cancer patients to facilitate or guide therapeutic decision making.

Other protein microarrays include forward protein microarrays (PMAs) and antibody microarrays (AMAs). PMAs immobilize individual purified and sometimes denatured recombinant proteins on the microarray that are screened by antibodies and other small compounds. AMAs immobilize antibodies that capture analytes from the sample applied on the microarray. The target protein is detected either by direct labeling or a secondary labeled antibody against a different epitope on the analyte target protein (sandwich approach). Both PMAs and AMAs can be classified as forward phase arrays as they involve immobilization of a bait to capture an analyte. In forward phase arrays, each array is incubated with one test sample such as a cellular lysate or a patient's serum, but multiple analytes in the sample are tested simultaneously.

Experimental Design and Procedure

Depending on the research question or the type and aim of the study, RPMA can be designed by selecting the content of the array, the number of samples, sample placement within micro-plates, array layout, type of microarrayer, correct detection antibody, signal detection method, inclusion of control and quality control of the samples. The actual experiment is then set up in the laboratory and the results obtained are quantified and analyzed. The experimental stages are listed below:

Sample Collection

Cells are grown in T-25 flasks at 37 degree and 5% CO_2 in appropriate medium. Depending on the design of the study, after cells are confluent they could be treated with drugs, growth factors or they could be irradiated before lysis step. For time course studies, a stimulant is added to a set of flasks concurrently and the flasks are then processed at different time points. For drug dose studies, a set of flasks are treated with different doses of the drug and all the flasks are collected at the same time.

If a RPMA containing cell fraction lysates of a tissue/s is to be made, laser capture microdissection (LCM) or fine needle aspiration methods is used to isolate specific cells from a region of tissue microscopically.

Cell Lysis

Pellets from cells collected through any of the above means are lysed with a cell lysis buffer to obtain high protein concentration.

Antibody Screening

Aliquots of the lysates are pooled and resolved by two-dimensional single lane SDS-PAGE followed by western blotting on a nitrocellulose membrane. The membrane is cut into four-millimeter strips, and each strip is probed with a different antibody. Strips with single band indicate specific antibodies that are suitable for RPMA use. Antibody performance should be also validated with a smaller sample size under identical condition before actual sample collection for RPMA.

RPMA Construction

Cell lysates are collected and are serially diluted six to ten times if using colorimetric techniques, or without dilution when fluorometric detection is used (due to the greater dynamic range of fluorescence than colorimetric detection). Serial dilutions are then plated in replicates into a 384- or a 1536-well microtiter plate. The lysates are then printed onto either nitrocellulose or PVDF membrane coated glass slides by a microarrayer such as Aushon BioSystem 2470 or Flexys robot (Genomic solution). Aushon 2470 with a solid pin system is the ideal choice as it can be used for producing arrays with very viscous lysates and it has humidity environmental control and automated slide supply system. That being said, there are published papers showing that Arrayit Microarray Printing Pins can also be used and produce microarrays with much higher throughput using less lysate. The membrane coated glass slides are commercially available from several different companies such as Schleicher and Schuell Bioscience, Grace BioLabs, Thermo Scientific, and SCHOTT Nexterion.

Immunochemical Signal Detection

After the slides are printed, non-specific binding sites on the array are blocked using a blocking buffer such as I-Block and the arrays are probed with a primary antibody followed by a secondary antibody. Detection is usually conducted with DakoCytomation catalyzed signal amplification (CSA) system. For signal amplification, slides are incubated with streptavidin-biotin-peroxidase complex followed by biotinyl-tyramide/hydrogen peroxide and streptavidin-peroxidase. Development is completed using hydrogen peroxide and scans of the slides are obtained (1). Tyramide signal amplification works as follows: immobilized horseradish peroxidase (HRP) converts tyramide into reactive intermediate in the presence of hydrogen peroxide. Activated tyramide binds to neighboring proteins close to a site where the activating HRP enzyme is bound. This leads to more tyramide molecule deposition at the site; hence the signal amplification.

Lance Liotta and Emanual Petricoin invented the RPMA technique in 2001, and have developed a multiplexed detection method using near-infrared fluorescent techniques. In this study, they report the use of a dual dye-based approach that can effectively double the number of endpoints observed per array, allowing, for example, both phospho-specific and total protein levels to be measured and analyzed at once.

Data Quantification and Analysis

Once immunostaining has been performed protein expression must then be quantified. The signal levels can be obtained by using the reflective mode of an ordinary optical flatbed scanner if a colorimetric detection technique is used or by laser scanning, such as with a TECAN LS system, if fluorescent techniques are used. Two programs available online (P-SCAN and ProteinScan) can then be used to convert the scanned image into numerical values. These programs quantify signal intensities at each spot and use a dose interpolation algorithm (DI_{25}) to compute a single normalized protein expression level value for each sample. Normalization is necessary to account for differences in total protein concentration between each sample and so that antibody staining can be directly compared between samples. This can be achieved by performing an experiment in parallel in which total proteins are stained by colloidal gold total protein staining or Sypro Ruby total protein staining. When multiple RPMAs are analyzed, the signal intensity values can be displayed as a heat map, allowing for Bayesian clustering analysis and profiling of signaling pathways. An optimal software tool, custom designed for RPMAs is called Microvigene, by Vigene Tech, Inc.

Strengths

The greatest strength of RPMAs is that they allow for high throughput, multiplexed, ultra-sensitive detection of proteins from extremely small numbers of input material, a feat which cannot be done by conventional western blotting or ELISA. The small spot size on the microarray, ranging in diameter from 85 to 200 micrometres, enables the analysis of thousands of samples with the same antibody in one experiment. RPMAs have increased sensitivity and are capable of detecting proteins in the picogram range. Some researchers have even reported detection of proteins in the attogram range. This is a significant improvement over protein detection by ELISA, which requires microgram amounts of protein (6). The increase in sensitivity of RPMAs is due to the miniature format of the array, which leads to an increase in the signal density (signal intensity/area) coupled with tyramide deposition-enabled enhancement. The high sensitivity of RPMAs allows for the detection of low abundance proteins or biomarkers such as phosphorylated signaling proteins from very small amounts of starting material such as biopsy samples, which are often contaminated with normal tissue. Using laser capture microdissection lysates can be analyzed from as few as 10 cells, with each spot containing less than a hundredth of a cell equivalent of protein.

A great improvement of RPMAs over traditional forward phase protein arrays is a reduction in the number of antibodies needed to detect a protein. Forward phase protein arrays typically use a sandwich method to capture and detect the desired protein. This implies that there must be two epitopes on the protein (one to capture the protein and one to detect the protein) for which specific antibodies are available. Other forward phase protein microarrays directly label the samples, however there is often variability in the labeling efficiency for different protein, and often the labeling destroys the epitope to which the antibody binds. This problem is overcome by RPMAs as sample need not be labeled directly.

Another strength of RPMAs over forward phase protein microarrays and western blotting is the uniformity of results, as all samples on the chip are probed with the same primary and secondary antibody and the same concentration of amplification reagents for the same length of time. This allows for the quantification of differences in protein levels across all samples. Furthermore, printing each sample, on the chip in serial dilution (colorimetric) provides an internal control to ensure analysis is performed only in the linear dynamic range of the assay. Optimally, printing of calibrators and high and low controls directly on the same chip will then provide for unmatched ability to quantitatively measure each protein over time and between experiments. A problem that is encountered with tissue microarrays is antigen retrieval and the inherent subjectivity of immunohistochemistry. Antibodies, especially phospho-specific reagents, often detect linear peptide sequences that may be masked due to the three-dimensional conformation of the protein. This problem is overcome with RPMAs as the samples can be denatured, revealing any concealed epitopes.

Weaknesses

The biggest limitation of RPMA, as is the case for all immunoassays, is its dependence on antibodies for detection of proteins. Currently there is a limited but rapidly growing number of signaling proteins for which antibodies exist that give an analyzable signal. In addition, finding the appropriate antibody could require extensive screening of many antibodies by western blotting prior to beginning RPMA analysis. To overcome this issue, two open resource databases have been created to display western blot results for antibodies that have good binding specificity within the expected range. Furthermore, RPMAs, unlike western blots, do not resolve protein fractions by molecular weight. Thus, it is critical that upfront antibody validation be performed.

History

RPMA was first introduced in 2001 in a paper by Lance Liotta and Emanuel Petricoin who invented the technology. The authors used the technique to successfully analyze the state of pro-survival checkpoint protein at the microscopic transition stage using laser capture microdissection of histologically normal prostate epithelium, prostate intraepithelial neoplasia, and patient-matched invasive prostate cancer. Since then RPMA has been used in many basic biology, translational and clinical research. In addition, the technique has now been brought into clinical trials for the first time whereby patients with metastatic colorectal and breast cancers are chosen for therapy based on the results of the RPMA. This technique has been commercialized for personalized medicine-based applications by Theranostics Health, Inc.

Principle

Reverse phase protein microarrays employ two existing technologies; (i) laser capture micro dissection (LCM), where stained tissue slide is placed under a microscope and the tissue is visualized in real-time manner, and (ii) microarray fabrication (Paweletz CP et al 2001). In RPPA whole protein lysate either from histopathologically relevant cell populations from diseased tissue procured by LCM or fine-needle aspiration cytology (FNAC), from cultured cells, serum, body fluid or lumps and masses is immobilized on coated slides. Signal intensity depends on amount of analyte protein present in individual spot thus a range of serial dilutions of each cell lysate are printed/immobilized on slide to make sure that analyte of interest remains within the linear range of detection and this dilution series is used to estimate expression of a protein. Then it is probed with highly specific antibody followed by detection with a secondary antibody labeled usually with fluorescent dyes like Cy3 or Cy5. Binding affinity of primary antibody and saturation effect is another important parameter to be taken care of; therefore highly specific antibody is used in analysis.

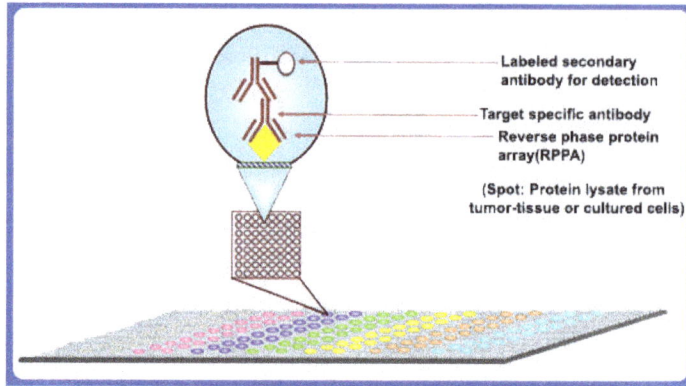

Reverse phase protein array (RPPA)

Proteins from tumor or cultured cells are immobilized onto the glass slide. The antibodies, in the solution that is used for evaluation, bind to the corresponding proteins on the microarray slide surface. This interaction can be detected by using a labeled secondary antibody, which has affinity to the primary antibody.

Procedure

- Tissue sample collection/ cell culture sample/ serum/ other body fluid

- Laser capture micro-dissection (LCM) for tissue

- Protein extraction/protein lysates

- Serial dilution of protein lysates

- Printing on slide

- Incubation with specific antibody

- Detection

  ○ Fluorescence

  ○ Enzyme-activated colorimetry

  ○ Chemiluminescence

- Analysis of reverse phase protein array data

Applications

The major applications of RPPAs include biomarker discovery and signaling pathway profiling. Srivastava et al (Srivastava M et al 2006) successfully used RPPA platform for validation of plausible biomarker for cystic fibrosis. An antibody microarray platform which had 507-features was used at first to identify and quantify low abundance signaling protein in serum from cystic fibrosis patients, and several candidates were identified. Further, using RPPA platform three of the antibodies, prostaglandin E2 Synthase (p23), glucocorticoid receptor and Caspase 4 were quantified and recognized as serum proteomic signature for cystic fibrosis.

Similarly, RPPA platform is also useful to profile proteins involved in disease progression. RPPA was introduced by Paweletz et al. where whole protein lysates from histo-pathologically relevant cell populations from human tissue procured by LCM were immobilized on nitrocellulose coated slides. Various stages of microscopic progressing cancer lesions were captured within individual patients on array and each patient set was arrayed in several dilutions (Paweletz CP et al 2001). Cancer progression was found to be related to increased phosphorylation of Akt, suppression of apoptosis signaling, and decreased ERK phosphorylation.

RPPAs can also be used for quantitative profiling of disease associated proteins or molecular networks profiling (Wulfkuhle JD et al 2003; Grubb RL et al 2003; Sheehan KM et al 2005), studying therapeutic responses (Carey MS et al 2010), validation of mass spectrometry discovered candidate biomarkers by RPPA (Van-Meter AJ et al 2008; Mueller C et al 2010) and personalized medicine (Mueller C et al 2010).

Advantages and Challenges

There are several advantages of using RPPAs for detecting protein-protein interaction. One can perform multiplex analysis where in multiple analytes could be evaluated simultaneously from relatively smaller numbers of cells than required by other techniques. RPPAs can be used for studying post-translational modifications (PTMs), such as phosphorylation and de-phosphorylation mediated by protein kinases. These PTMs are mediated by protein kinases which are critical in transduction networks. The assays performed using RPPAs provide high sensitivity in comparison to western blot (detection capabilities of 50 fg/l, or 1,000 to 5,000 molecules per spot (Liotta et al. 2003; Paweletz et al. 2001; Geho D et al. 2005). Apart from high sensitivity, RPPAs also provide high reproducibility and robustness as it does not require any direct labelling of the sample (Sheehan KM et al 2005). Problems associated with antigen retrieval while using other types of microarrays can be avoided by using RPPA, since RPPAs can use denatured lysates.

Despite having distinct advantages, RPPAs have certain drawbacks, which need to be addressed. Availability and specificity of primary antibodies are one of the challenges of RPPAs, which limits the use of this technique. Other problems associated with RPPAs are sample degradation, preservation and proper data normalization.

The RPPAs provides a high-throughput proteomic platform with high sensitivity. With continuous advancement in technique its wide applications has been reported in various fields including clinical research. Although there are several challenges to be taken care of, but this promising approach may be helpful for biomarker discovery & validation and personalized medicine.

Cell-free Protein Array

Cell-free protein array technology produces protein microarrays by performing in vitro synthesis of the target proteins from their DNA templates. This method of synthesizing protein microarrays overcomes the many obstacles and challenges faced by tradition- al methods of protein array production that have prevented widespread adoption of protein microarrays in proteomics. Protein arrays made from this technology can be used for testing protein–protein interactions, as well as protein interactions with other cellular molecules such as DNA and lipids. Other applications include enzymatic inhi- bition assays and screenings of antibody specificity.

Overview / Background

The runaway success of DNA microarrays has generated much enthusiasm for pro- tein microarrays. However, protein microarrays have not quite taken off as expected, even with the necessary tools and know-how from DNA microarrays being in place and ready for adaptation. One major reason is that protein microarrays are much more la- borious and technically challenging to construct than DNA microarrays.

The traditional methods of producing protein arrays require the separate *in vivo* ex- pression of hundreds or thousands of proteins, followed by separate purification and immobilization of the proteins on a solid surface. Cell-free protein array technology at- tempts to simplify protein microarray construction by bypassing the need to express the proteins in bacteria cells and the subsequent need to purify them. It takes advantage of available cell-free protein synthesis technology which has demonstrated that protein synthesis can occur without an intact cell as long as cell extracts containing the DNA tem- plate, transcription and translation raw materials and machinery are provided. Common sources of cell extracts used in cell-free protein array technology include wheat germ, *Escherichia coli*, and rabbit reticulocyte. Cell extracts from other sources such as hyper- thermophiles, hybridomas, Xenopusoocytes, insect, mammalian and human cells have also been used.

The target proteins are synthesized *in situ* on the protein microarray, directly from the DNA template, thus skipping many of the steps in traditional protein microarray pro- duction and their accompanying technical limitations. More importantly, the expres- sion of the proteins can be done in parallel, meaning all the proteins can be expressed together in a single reaction. This ability to multiplex protein expression is a major time-saver in the production process.

Methods of Synthesis

In Situ methods

In the *in situ* method, protein synthesis is carried out on a protein array surface that is pre-coated with a protein-capturing reagent or antibody. Once the newly synthesized proteins are released from the ribosome, the tag sequence that is also synthesized at the N- or C-terminus of each nascent protein will be bound by the capture reagent or antibody, thus immobilizing the proteins to form an array. Commonly used tags include polyhistidine (His)6 and glutathione s-transferase (GST).

Various research groups have developed their own methods, each differing in their approach, but can be summarized into 3 main groups.

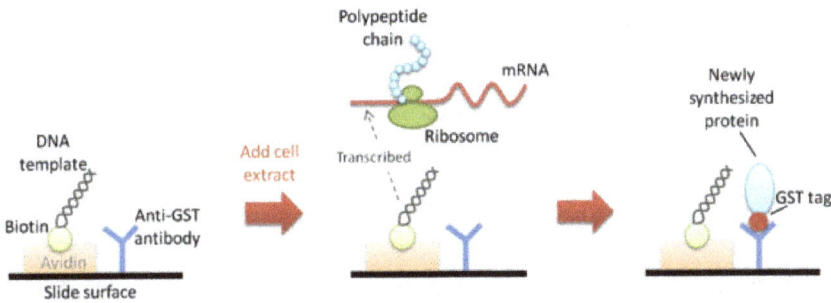

Diagram of NAPPA

Nucleic Acid Programmable Protein Array (NAPPA)

NAPPA uses DNA template that has already been immobilized onto the same protein capture surface. The DNA template is biotinylated and is bound to avidin that is pre-coated onto the protein capture surface. Newly synthesized proteins which are tagged with GST are then immobilized next to the template DNA by binding to the adjacent polyclonal anti-GST capture antibody that is also pre-coated onto the capture surface. The main drawback of this method is the extra and tedious preparation steps at the beginning of the process: (1) the cloning of cDNAs in an expression-ready vector; and (2) the need to biotinylate the plasmid DNA but not to interfere with transcription. Moreover, the resulting protein array is not 'pure' because the proteins are co-localized with their DNA templates and capture antibodies.

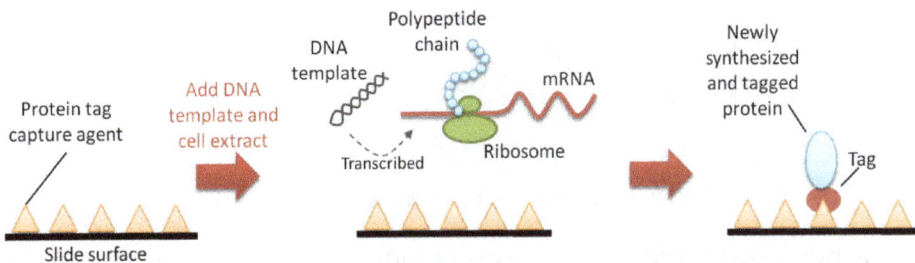

Diagram of PISA

Protein *in Situ* Array (PISA)

Unlike NAPPA, PISA completely bypasses DNA immobilization as the DNA template is added as a free molecule in the reaction mixture. In 2006, another group refined and miniaturized this method by using multiple spotting technique to spot the DNA template and cell-free transcription and translation mixture on a high-density protein microarray with up to 13,000 spots. This was made possible by the automated system used to accurately and sequentially supply the reagents for the transcription/translation reaction occurs in a small, sub-nanolitre droplet.

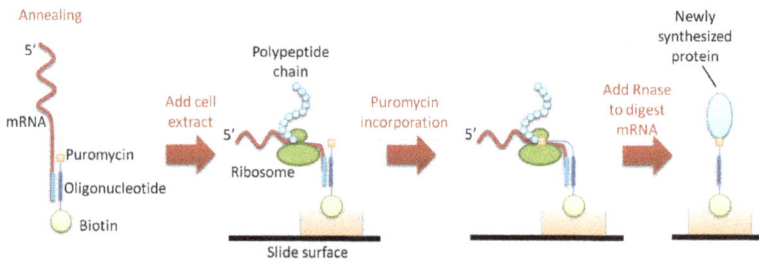

Diagram of *In Situ* Puromycin-capture

In Situ Puromycin-capture

This method is an adaptation of mRNA display technology. PCR DNA is first transcribed to mRNA, and a single-stranded DNA oligonucleotide modified with biotin and puromycin on each end is then hybridized to the 3'-end of the mRNA. The mRNAs are then arrayed on a slide and immobilized by the binding of biotin to streptavidin that is pre-coated on the slide. Cell extract is then dispensed on the slide for *in situ* translation to take place. When the ribosome reaches the hybridized oligonucleotide, it stalls and incorporates the puromycin molecule to the nascent polypeptide chain, thereby attaching the newly synthesized protein to the microarray via the DNA oligonucleotide. A pure protein array is obtained after the mRNA is digested with RNase. The protein spots generated by this method are very sharply defined and can be produced at a high density.

Nano-well Array Format

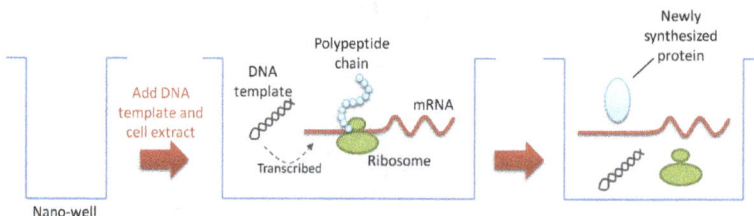

Schematic diagram of the nano-well array format

Nanowell array formats are used to express individual proteins in small volume reaction vessels or nanowells. This format is sometimes preferred because it avoids the need to immobilize the target protein which might result in the potential loss of protein activity. The miniaturization of the array also conserves solution and precious compounds that might be used in screening assays. Moreover, the structural properties of individual wells help to prevent cross-contamination among chambers. In 2012 an improved NAPPA was published, which used a nanowell array to prevent diffusion. Here the DNA was immobilized in the well together with an anti-GST antibody. Then cell-free expression mix was added and the wells closed by a lid. The nascent proteins containing a GST-tag were bound to the well surface enabling a NAPPA-array with higher density and nearly no cross-contaminations.

DNA Array to Protein Array (DAPA)

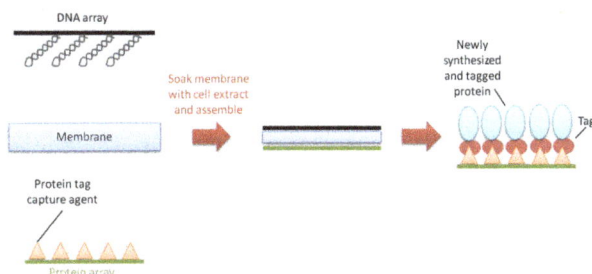

Schematic diagram of DAPA

DNA array to protein array (DAPA) is a method developed in 2007 to repeatedly produce protein arrays by 'printing' them from a single DNA template array, on demand. It starts with the spotting and immobilization of an array of DNA templates onto a glass slide. The slide is then assembled face-to-face with a second slide pre-coated with a protein-capturing reagent, and a membrane soaked with cell extract is placed between the two slides for transcription and translation to take place. The newly synthesized his-tagged proteins are then immobilized onto the slide to form the array. In the publication in 18 of 20 replications a protein microarray copy could be generated. Potentially the process can be repeated as often as needed, as long as the DNA is unharmed by DNAses, degradation or mechanical abrasion.

Advantages

Many of the advantages of cell-free protein array technology address the limitations of cell-based expression system used in traditional methods of protein microarray production.

Rapid and Cost-effective

The method avoids DNA cloning (with the exception of NAPPA) and can quickly convert genetic information into functional proteins by using PCR DNA. The reduced steps in production and the ability to miniaturize the system saves on reagent consumption and cuts production costs.

Improves Protein Availability

Many proteins, including antibodies, are difficult to express in host cells due to problems with insolubility, disulfide bonds or host cell toxicity. Cell-free protein array makes many of such proteins available for use in protein microarrays.

Enables Long Term Storage

Unlike DNA, which is a highly stable molecule, proteins are a heterogeneous class of molecules with different stability and physiochemical properties. Maintaining the proteins' folding and function in an immobilized state over long periods of storage is a major challenge for protein microarrays. Cell-free methods provide the option to quickly obtaining protein microarrays on demand, thus eliminating any problems associated with long-term storage.

Flexible

The method is amenable to a range of different templates: PCR products, plasmids and mRNA. Additional components can be included during synthesis to adjust the environment for protein folding, disulfide bond formation, modification or protein activity.

Limitation

- Post-translational modification of proteins in proteins generated by cell-free protein synthesis is still limited compared to the traditional methods, and may not be as biologically relevant.

Applications

Protein interactions: To screen for protein–protein interactions and protein interactions with other molecules such as metabolites, lipids, DNA and small molecules.; enzyme inhibition assay: for high throughput drug candidate screening and to discover novel enzymes for use in biotechnology; screening antibody specificity.

Multiple Spotting Technique (MIST)

Another method of making a protein array is the multiple spotting technique (MIST), which was developed by Angenendt et al. (Angenendt et al. 2004). Here, the support-like a slide is pre-coated with a protein capture agent. The first spot printed on slide consists of the DNA template. The second spot printed contains the in vitro translation mixture, which is added exactly on top of the spot containing the DNA template. The translation of the proteins takes place, and protein-capture agent captures newly formed protein on slide, and an array is constructed.

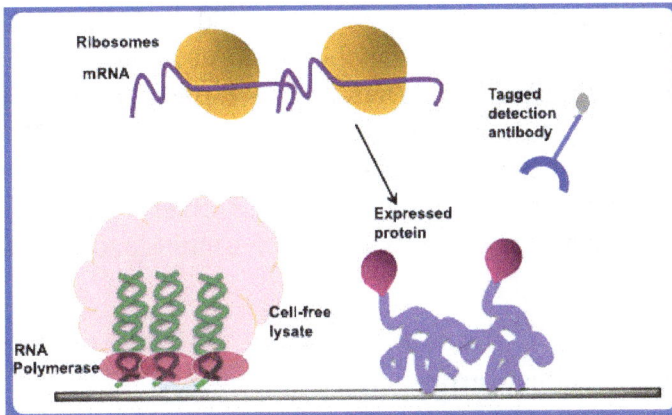

Multiple spotting technique (MIST).

First spotting step involves the spotting of DNA template coding for the protein of interest in as little as fg quantities, onto the solid array surface. Second spotting: After the template DNA is spotted on to the array surface, the cell-free lysate is then transferred exactly on top of the first spot. This second spotting step marks the beginning of expression of the template DNA. The proteins produced by cell-free expression from the corresponding DNA templates are immobilized on the array surface either through a tag-capturing agent or more commonly, by means of non-specific interactions. These proteins can then be detected by suitably tagged antibodies.

Merits

- Unpurified DNA products used as template
- Very high-density protein arrays generated

Demerits

- Loss of signal intensity with prolonged incubation time
- Non-specific protein binding
- Time consuming process

DNA Array to Protein Array (DAPA)

This concept was developed by He et al. in 2008 (He et al. 2008). In the DAPA concept, two different slides are used. One of the slides coated with Ni-NTA has the PCR-amplified DNA fragments, which encode the protein of interest fused with a tag immobilized onto it; while the other Ni-NTA slide has the protein-tag capturing agent. The slides are placed face-to-face and a permeable membrane is kept between them. The permeable membrane has the cell-free lysate, and thus protein expression is initiated in between the two slides.

The newly synthesized proteins are produced on the slide with the DNA template, which then penetrate the membrane and get immobilized on surface of the slide bearing the tag capture reagent. Thus, a replica of the DNA array is formed on the capture slide. DAPA leads to the construction of a pure protein array i.e. an array with no DNA contamination, as the two are kept separate throughout the experiment. Another important advantage of DAPA is that the DNA template slide is reusable i.e. the proteins can be printed many times from the same DNA, making the process less time-consuming and cost-effective.

Merits

- Reusable DNA template array

- Pure protein array generated

- DNA template array can be stored for long durations

Demerits

- Broadening of spots due to diffusion

- Not ascertained if multimeric proteins assemble effectively

- Time consuming process

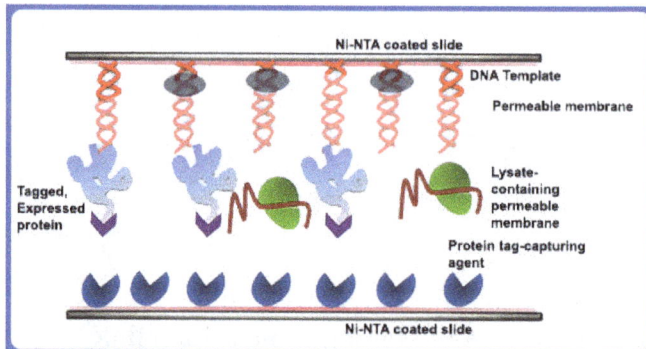

The microarray slide surface is coated with Nickel-nitrilotriacetic acid (Ni-NTA), which acts as a useful capture agent. PCR amplified DNA that codes for the protein of interest is immobilized on a Ni-NTA coated slide. A permeable membrane that is soaked with the cell-free extract is placed in between the immobilized DNA template slide and a slide having the protein tag-capturing agent. The newly expressed proteins penetrate the membrane and bind to the protein purification slide.

Halo-link Protein Array

This is an amalgamation of technologies developed by Promega and is useful in generating tightly immobilized arrays. The Halolink array consists of a DNA construct encoding the gene of interest fused with the 'Halotag', a 33 KDa mutated bacterial hydrolase. The protein is constructed using a cell free expression system (WGE or RRL). The newly formed proteins are captured on a polyethylene glycol coated glass slide, which has been activated by HaloTag ligands. The HaloTag fused to the protein of interest binds the Halo ligand on the slide by covalent bonding, thus enabling capture of the desired protein. The strong bond afforded by the covalent linkage prevents material loss during washing, which is always a concern for any microarray. It also allows oriented capture of proteins, hence keeping the protein activity unaffected.

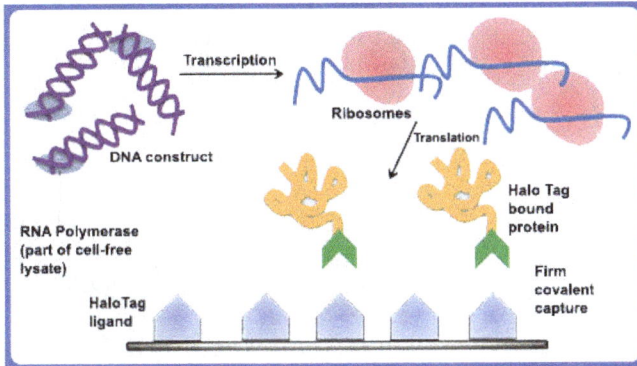

The template DNA coding for the protein of interest along with the HaloTag. The mode of interaction between HaloTag and its ligand is through covalent bonding, thereby ensuring firm capture of the protein on the array surface without any material loss during washing.

Merits

- Strong covalent bond between protein and ligand

- No material loss during washing

- Oriented capture of protein

- No non-specific adsorption

- Easy quantification

- No need for a microarrayer printer

Demerits

- Possible loss of function on binding to Halotag

- HT application will require optimization of printing

Commonly Used Techniques for Cell-free Expression

Cell free protein expression is the expression of proteins from template DNA, which is inserted in plasmids or present as purified PCR products. The addition of a crude cell lysate containing the required cellular machinery for protein production such as enzymes, ribosomes, tRNA etc., and exogenously added co-factors - nucleotides,

ATP, salts, essential amino acids is required for in vitro transcription and translation (IVTT) from the gene of interest. Additionally, such a system, based on cell-free expression machinery used, may also allow protein folding and post-translational modifications to produce structurally and functionally mature protein. There are different types of cell-free expression systems. The source of DNA for cell-free protein expression is a DNA fragment, either PCR-produced or present in a plasmid vector, that contains all the genetic information promoter, translational initiator, gene of interest, translational terminator etc., in the correct orientation and reading frame for protein synthesis.

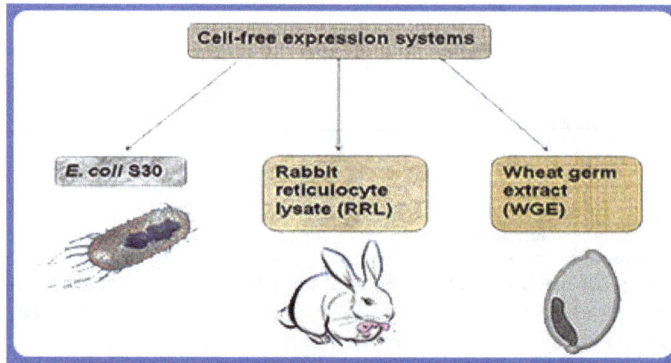

Different types of cell-free expression systems used for in vitro protein synthesis starting from DNA templates. These systems contain all the necessary components and machinery for transcription and translation. Some factors such as energy generating components, essential amino acids etc. need to be added to the system for successful protein synthesis.

Escherichia Coli

Escherichia coli is a gram-negative, facultatively anaerobic, rod-shaped, coliform bacterium of the genus *Escherichia* that is commonly found in the lower intestine of warm-blooded organisms (endotherms). Most *E. coli* strains are harmless, but some serotypes can cause serious food poisoning in their hosts, and are occasionally responsible for product recalls due to food contamination. The harmless strains are part of the normal flora of the gut, and can benefit their hosts by producing vitamin K_2, and preventing colonization of the intestine with pathogenic bacteria, having a symbiotic relationship. *E. coli* is expelled into the environment within fecal matter. The bacterium grows massively in fresh fecal matter under aerobic conditions for 3 days, but its numbers decline slowly afterwards.

E. coli and other facultative anaerobes constitute about 0.1% of gut flora, and fecal–oral transmission is the major route through which pathogenic strains of the bacterium cause disease. Cells are able to survive outside the body for a limited amount of time, which makes them potential indicator organisms to test environmental samples for fecal contamination. A growing body of research, though, has examined environmentally persistent *E. coli* which can survive for extended periods outside of a host.

The bacterium can be grown and cultured easily and inexpensively in a laboratory setting, and has been intensively investigated for over 60 years. *E. coli* is a chemoheterotroph whose chemically defined medium must include a source of carbon and energy.*E. coli* is the most widely studied prokaryoticmodel organism, and an important species in the fields of biotechnology and microbiology, where it has served as the host organism for the majority of work with recombinant DNA. Under favorable conditions, it takes only 20 minutes to reproduce.

Biology and Biochemistry

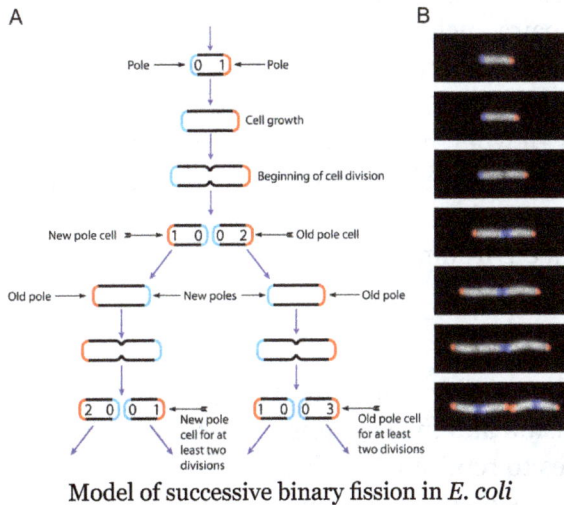

Model of successive binary fission in *E. coli*

Type and Morphology

E. coli is a Gram-negative, facultative anaerobic (that makes ATP by aerobic respiration if oxygen is present, but is capable of switching to fermentation or anaerobic respiration if oxygen is absent) and nonsporulating bacterium. Cells are typically rod-shaped, and are about 2.0 μm long and 0.25–1.0 μm in diameter, with a cell volume of 0.6–0.7 μm^3.

E. coli stains Gram-negative because its cell wall is composed of a thin peptidoglycan layer and an outer membrane. During the staining process, *E. coli* picks up the color of the counterstain safranin and stains pink. The outer membrane surrounding the cell wall provides a barrier to certain antibiotics such that *E. coli* is not damaged by penicillin.

Strains that possess flagella are motile. The flagella have a peritrichous arrangement.

Metabolism

E. coli can live on a wide variety of substrates and uses mixed-acid fermentation in anaerobic conditions, producing lactate, succinate, ethanol, acetate, and carbon dioxide. Since many pathways in mixed-acid fermentation produce hydrogen gas, these pathways require the levels of hydrogen to be low, as is the case when *E. coli* lives together with hydrogen-consuming organisms, such as methanogens or sulphate-reducing bacteria.

Culture Growth

Optimum growth of *E. coli* occurs at 37 °C (98.6 °F), but some laboratory strains can multiply at temperatures up to 49 °C (120 °F).*E. coli* grows in a variety of defined laboratory media, such as lysogeny broth, or any medium that contains glucose, ammonium phosphate, monobasic, sodium chloride, magnesium sulfate, potassium phosphate, dibasic, and water. Growth can be driven by aerobic or anaerobic respiration, using a large variety of redox pairs, including the oxidation of pyruvic acid, formic acid, hydrogen, and amino acids, and the reduction of substrates such as oxygen, nitrate, fumarate, dimethyl sulfoxide, and trimethylamine N-oxide.*E. coli* is classified as a facultative anaerobe. It uses oxygen when it is present and available. It can, however, continue to grow in the absence of oxygen using fermentation or anaerobic respiration. The ability to continue growing in the absence of oxygen is an advantage to bacteria because their survival is increased in environments where water predominates.

Cell Cycle

The bacterial cell cycle is divided into three stages. The B period occurs between the completion of cell division and the beginning of DNA replication. The C period encompasses the time it takes to replicate the chromosomal DNA. The D period refers to the stage between the conclusion of DNA replication and the end of cell division. The doubling rate of *E. coli* is higher when more nutrients are available. However, the length of the C and D periods do not change, even when the doubling time becomes less than the sum of the C and D periods. At the fastest growth rates, replication begins before the previous round of replication has completed, resulting in multiple replication forks along the DNA and overlapping cell cycles.

Unlike eukaryotes, prokaryotes do not rely upon either changes in gene expression or changes in protein synthesis to control the cell cycle. This probably explains why they do not have similar proteins to those used by eukaryotes to control their cell cycle, such as cdk1. This has led to research on what the control mechanism is in prokaryotes. Recent evidence suggests that it may be membrane- or lipid-based.

Genetic Adaptation

E. coli and related bacteria possess the ability to transfer DNA via bacterial conjugation or transduction, which allows genetic material to spread horizontally through an existing population. The process of transduction, which uses the bacterial virus called a bacteriophage, is where the spread of the gene encoding for the Shiga toxin from the *Shigella* bacteria to *E. coli* helped produce *E. coli* O157:H7, the Shiga toxin-producing strain of *E. coli*.

Diversity

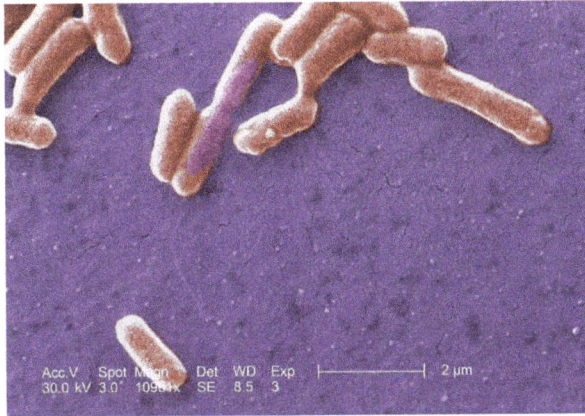

Scanning electron micrograph of an *E. coli* colony.

E. coli encompasses an enormous population of bacteria that exhibit a very high degree of both genetic and phenotypic diversity. Genome sequencing of a large number of isolates of *E. coli* and related bacteria shows that a taxonomic reclassification would be desirable. However, this has not been done, largely due to its medical importance, and *E. coli* remains one of the most diverse bacterial species: only 20% of the genes in a typical *E. coli* genome is shared among all strains.

In fact, from the evolutionary point of view, the members of genus *Shigella* (*S. dysenteriae*, *S. flexneri*, *S. boydii*, and *S. sonnei*) should be classified as *E. coli* strains, a phenomenon termed taxa in disguise. Similarly, other strains of *E. coli* (e.g. the K-12 strain commonly used in recombinant DNA work) are sufficiently different that they would merit reclassification.

A strain is a subgroup within the species that has unique characteristics that distinguish it from other strains. These differences are often detectable only at the molecular level; however, they may result in changes to the physiology or lifecycle of the bacterium. For example, a strain may gain pathogenic capacity, the ability to use a unique carbon source, the ability to take upon a particular ecological niche, or the ability to resist antimicrobial agents. Different strains of *E. coli* are often host-specific, making it possible to determine the source of fecal contamination in environmental samples. For example, knowing which *E. coli* strains are present in a water sample allows researchers to make assumptions about whether the contamination originated from a human, another mammal, or a bird.

Serotypes

A common subdivision system of *E. coli*, but not based on evolutionary relatedness, is by serotype, which is based on major surface antigens (O antigen: part of lipopolysaccharide layer; H: flagellin; K antigen: capsule), e.g. O157:H7). It is, however, common to cite only the serogroup, i.e. the O-antigen. At present, about 190 serogroups are known. The common laboratory strain has a mutation that prevents the formation of an O-antigen and is thus not typeable.

Genome Plasticity and Evolution

Like all lifeforms, new strains of *E. coli* evolve through the natural biological processes of mutation, gene duplication, and horizontal gene transfer; in particular, 18% of the genome of the laboratory strain MG1655 was horizontally acquired since the divergence from *Salmonella*. *E. coli* K-12 and *E. coli* B strains are the most frequently used varieties for laboratory purposes. Some strains develop traits that can be harmful to a host animal. These virulent strains typically cause a bout of diarrhea that is often self-limiting in healthy adults but is frequently lethal to children in the developing world. More virulent strains, such as O157:H7, cause serious illness or death in the elderly, the very young, or the immunocompromised.

The genera *Escherichia* and *Salmonella* diverged around 102 million years ago (credibility interval: 57–176 mya) which coincides with the divergence of their hosts: the former being found in mammals and the latter in birds and reptiles. This was followed by a split of an *Escherichia* ancestor into five species (*E. albertii*, *E. coli*, *E. fergusonii*, *E. hermannii*, and *E. vulneris*). The last *E. coli* ancestor split between 20 and 30 million years ago.

The long-term evolution experiments using *E. coli*, begun by Richard Lenski in 1988, have allowed direct observation of major evolutionary shifts in the laboratory. In this experiment, one population of *E. coli* unexpectedly evolved the ability to aerobically metabolize citrate, which is extremely rare in *E. coli*. As the inability to grow aerobically is normally used as a diagnostic criterion with which to differentiate *E. coli* from other, closely related bacteria, such as *Salmonella*, this innovation may mark a speciation event observed in the laboratory.

Neotype Strain

E. coli is the type species of the genus (*Escherichia*) and in turn *Escherichia* is the type genus of the family Enterobacteriaceae, where the family name does not stem from the genus *Enterobacter* + "i" (sic.) + "aceae", but from "enterobacterium" + "aceae" (enterobacterium being not a genus, but an alternative trivial name to enteric bacterium).

The original strain described by Escherich is believed to be lost, consequently a new type strain (neotype) was chosen as a representative: the neotype strain is U5/41^T, also known under the deposit names DSM 30083, ATCC 11775, and NCTC 9001, which is pathogenic to chickens and has an O1:K1:H7 serotype. However, in most studies, either O157:H7, K-12 MG1655, or K-12 W3110 were used as a representative *E. coli*. The genome of the type strain has only lately been sequenced.

Phylogeny of *E. Colistrains*

A large number of strains belonging to this species have been isolated and characterised. In addition to serotype (*vide supra*), they can be classified according to their phylogeny, i.e. the inferred evolutionary history, as shown below where the species is divided into six groups. Particularly the use of whole genome sequences yields highly supported phylogenies. Based on such data, five subspecies of *E. coli* were distinguished.

The link between phylogenetic distance ("relatedness") and pathology is small,*e.g.* the O157:H7 serotype strains, which form a clade ("an exclusive group")—group E below—are all enterohaemorrhagic strains (EHEC), but not all EHEC strains are closely related. In fact, four different species of *Shigella* are nested among *E. coli* strains (*vide supra*), while *E. albertii* and *E. fergusonii* are outside of this group. Indeed, all *Shigella* species were placed within a single subspecies of *E. coli* in a phylogenomic study that included the type strain,and for this reason an according reclassification is difficult. All commonly used research strains of *E. coli* belong to group A and are derived mainly from Clifton's K-12 strain (λ^+ F^+; O16) and to a lesser degree from d'Herelle's *Bacillus coli* strain (B strain)(O7).

Genomics

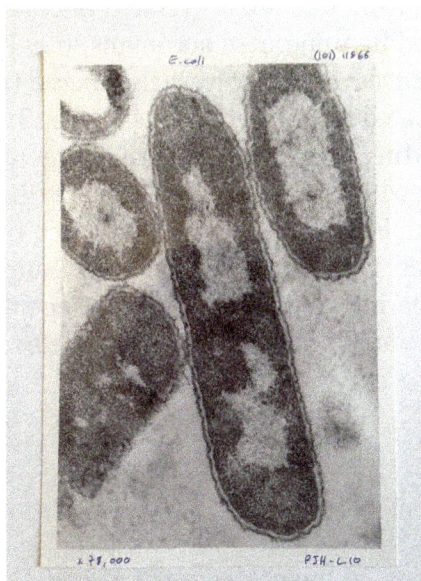

Early electron microscopy

The first complete DNA sequence of an *E. coligenome* (laboratory strain K-12 derivative MG1655) was published in 1997. It was found to be a circular DNA molecule 4.6 million base pairs in length, containing 4288 annotated protein-coding genes (organized into 2584 operons), seven ribosomal RNA (rRNA) operons, and 86 transfer RNA (tRNA) genes. Despite having been the subject of intensive genetic analysis for about 40 years, a large number of these genes were previously unknown. The coding density was found to be very high, with a mean distance between genes of only 118 base pairs. The genome was observed to contain a significant number of transposable genetic elements, repeat elements, cryptic prophages, and bacteriophage remnants.

Today, several hundred complete genomic sequences of *Escherichia* and *Shigella* species are available. The genome sequence of the type strain of *E. coli* has been added to this collection not before 2014. Comparison of these sequences shows a remarkable amount of diversity; only about 20% of each genome represents sequences present in every one of the isolates, while around 80% of each genome can vary among isolates. Each individual genome contains between 4,000 and 5,500 genes, but the total number of different genes among all of the sequenced *E. coli* strains (the pangenome) exceeds 16,000. This very large variety of component genes has been interpreted to mean that two-thirds of the *E. coli* pangenome originated in other species and arrived through the process of horizontal gene transfer.

Gene Nomenclature

Genes in *E. coli* are usually named by 4-letter acronyms that derive from their function (when known) and italicized. For instance, *recA* is named after its role in homologous recombination plus the letter A. Functionally related genes are named *recB*, *recC*, *recD* etc. The proteins are named by uppercase acronyms, e.g. RecA, RecB, etc. When the genome of *E. coli* was sequenced, all genes were numbered (more or less) in their order on the genome and abbreviated by b numbers, such as b2819 (= *recD*). The "b" names were created after Fred Blattner who led the genome sequence effort. Another numbering system was introduced with the sequence of another *E. coli* strain, W3110, which was sequenced in Japan and hence uses numbers starting by JW... (Japanese W3110), e.g. JW2787 (= *recD*). Hence, *recD* = b2819 = JW2787. Note, however, that most databases have their own numbering system, e.g. the EcoGene database uses EG10826 for *recD*. Finally, ECK numbers are specifically used for alleles in the MG1655 strain of *E. coli* K-12.

Proteomics

Proteome

Several studies have investigated the proteome of *E. coli*. By 2006, 1,627 (38%) of the 4,237 open reading frames (ORFs) had been identified experimentally.

Interactome

The interactome of *E. coli* has been studied by affinity purification and mass spectrometry (AP/MS) and by analyzing the binary interactions among its proteins.

Protein complexes. A 2006 study purified 4,339 proteins from cultures of strain K-12 and found interacting partners for 2,667 proteins, many of which had unknown functions at the time. A 2009 study found 5,993 interactions between proteins of the same *E. coli* strain, though these data showed little overlap with those of the 2006 publication.

Binary interactions. Rajagopala *et al.* (2014) have carried out systematic yeast two-hybrid screens with most *E. coli* proteins, and found a total of 2,234 protein-protein interactions. This study also integrated genetic interactions and protein structures and mapped 458 interactions within 227 protein complexes.

Normal Microbiota

E. coli belongs to a group of bacteria informally known as coliforms that are found in the gastrointestinal tract of warm-blooded animals.*E. coli* normally colonizes an infant's gastrointestinal tract within 40 hours of birth, arriving with food or water or from the individuals handling the child. In the bowel, *E. coli* adheres to the mucus of the large intestine. It is the primary facultative anaerobe of the human gastrointestinal tract. (Facultative anaerobes are organisms that can grow in either the presence or absence of oxygen.) As long as these bacteria do not acquire genetic elements encoding for virulence factors, they remain benign commensals.

Therapeutic Use

Nonpathogenic *E. coli* strain Nissle 1917, also known as Mutaflor, and *E. coli* O83:K24:H31 (known as Colinfant) are used as probiotic agents in medicine, mainly for the treatment of various gastroenterological diseases, including inflammatory bowel disease.

Role in Disease

Most *E. coli* strains do not cause disease, but virulent strains can cause gastroenteritis, urinary tract infections, neonatalmeningitis, hemorrhagic colitis, and Crohn's disease. Common signs and symptoms include severe abdominal cramps, diarrhea, hemorrhagic colitis, vomiting, and sometimes fever. In rarer cases, virulent strains are also responsible for bowel necrosis (tissue death) and perforation without progressing to hemolytic-uremic syndrome, peritonitis, mastitis, septicemia, and gram-negative pneumonia. Very young children are more susceptible to develop severe illness, such as hemolytic uremic syndrome, however, healthy individuals of all ages are at risk to the severe consequences that may arise as a result of being infected with *E. coli*.

There is one strain, *E. coli*O157:H7, that produces the Shiga toxin (classified as a bioterrorism agent). This toxin causes premature destruction of the red blood cells, which then clog the body's filtering system, the kidneys, causing hemolytic-uremic syndrome (HUS). Signs of hemolytic uremic syndrome, include decreased frequency of urination, lethargy, and paleness of cheeks and inside the lower eyelids. In 25% of HUS patients, complications of nervous system occur, which in turn causes strokes due to small clots of blood which lodge in capillaries in the brain. This causes the body parts controlled by this region of the brain not to work properly. In addition, this strain causes the buildup of fluid (since the kidneys do not work), leading to edema around the lungs and legs and arms. This increase in fluid buildup especially around the lungs impedes the functioning of the heart, causing an increase in blood pressure.

Uropathogenic *E. coli* (UPEC) is one of the main causes of urinary tract infections. It is part of the normal flora in the gut and can be introduced in many ways. In particular for females, the direction of wiping after defecation (wiping back to front) can lead to fecal contamination of the urogenital orifices. Anal intercourse can also introduce this bacterium into the male urethra, and in switching from anal to vaginal intercourse, the male can also introduce UPEC to the female urogenital system. For more information, see the databases at the end of the article or UPEC pathogenicity.

In May 2011, one *E. coli* strain, O104:H4, was the subject of a bacterial outbreak that began in Germany. Certain strains of *E. coli* are a major cause of foodborne illness. The outbreak started when several people in Germany were infected with enterohemorrhagic *E. coli* (EHEC) bacteria, leading to hemolytic-uremic syndrome (HUS), a medical emergency that requires urgent treatment. The outbreak did not only concern Germany, but also 11 other countries, including regions in North America. On 30 June 2011, the German *Bundesinstitut für Risikobewertung (BfR)* (Federal Institute for Risk Assessment, a federal institute within the German Federal Ministry of Food, Agriculture and Consumer Protection) announced that seeds of fenugreek from Egypt were likely the cause of the EHEC outbreak.

Incubation Period

The time between ingesting the STEC bacteria and feeling sick is called the "incubation period". The incubation period is usually 3–4 days after the exposure, but may be as short as 1 day or as long as 10 days. The symptoms often begin slowly with mild belly pain or non-bloody diarrhea that worsens over several days. HUS, if it occurs, develops an average 7 days after the first symptoms, when the diarrhea is improving.

Treatment

The mainstay of treatment is the assessment of dehydration and replacement of fluid and electrolytes. Administration of antibiotics has been shown to shorten the course of illness and duration of excretion of enterotoxigenic *E. coli* (ETEC) in adults in endemic areas and in traveller's diarrhea, though the rate of resistance to commonly used antibiotics is increasing and they are generally not recommended. The antibiotic used depends upon susceptibility patterns in the particular geographical region. Currently, the antibiotics of choice are fluoroquinolones or azithromycin, with an emerging role for rifaximin. Oral rifaximin, a semisynthetic rifamycin derivative, is an effective and well-tolerated antibacterial for the management of adults with non-invasive traveller's diarrhea. Rifaximin was significantly more effective than placebo and no less effective than ciprofloxacin in reducing the duration of diarrhea. While rifaximin is effective in patients with *E. coli*-predominant traveller's diarrhea, it appears ineffective in patients infected with inflammatory or invasive enteropathogens.

Prevention

ETEC is the type of *E. coli* that most vaccine development efforts are focused on. Antibodies against the LT and major CFs of ETEC provide protection against LT-producing ETEC expressing homologous CFs. Oral inactivated vaccines consisting of toxin antigen and whole cells, i.e. the licensed recombinant cholera B subunit (rCTB)-WC cholera vaccine Dukoral have been developed. There are currently no licensed vaccines for ETEC, though several are in various stages of development. In different trials, the rCTB-WC cholera vaccine provided high (85–100%) short-term protection. An oral ETEC vaccine candidate consisting of rCTB and formalin inactivated *E. coli* bacteria expressing major CFs has been shown in clinical trials to be safe, immunogenic, and effective against severe diarrhoea in American travelers but not against ETEC diarrhoea in young children in Egypt. A modified ETEC vaccine consisting of recombinant *E. coli* strains over expressing the major CFs and a more LT-like hybrid toxoid called LCTBA, are undergoing clinical testing.

Other proven prevention methods for *E. coli* transmission include handwashing and improved sanitation and drinking water, as transmission occurs through fecal contamination of food and water supplies. Additionally, thoroughly cooking meat and avoiding consumption of raw, unpasteurized beverages, such as juices and milk are other proven methods for preventing E.coli. Lastly, avoid cross-contamination of utensils and work spaces when preparing food.

Causes and Risk Factors

- Working around livestock
- Consuming unpasteurized dairy product
- Eating undercooked meat
- Drinking impure water

Model Organism in Life Science Research
Role in Biotechnology

Because of its long history of laboratory culture and ease of manipulation, *E. coli* plays an important role in modern biological engineering and industrial microbiology. The work of Stanley Norman Cohen and Herbert Boyer in *E. coli*, using plasmids and restriction enzymes to create recombinant DNA, became a foundation of biotechnology.

E. coli is a very versatile host for the production of heterologousproteins, and various protein expression systems have been developed which allow the production of recombinant proteins in *E. coli*. Researchers can introduce genes into the microbes using plasmids which permit high level expression of protein, and such protein may be mass-produced in industrial fermentation processes. One of the first useful applications of recombinant DNA technology was the manipulation of *E. coli* to produce human insulin.

Many proteins previously thought difficult or impossible to be expressed in *E. coli* in folded form have been successfully expressed in *E. coli*. For example, proteins with multiple disulphide bonds may be produced in the periplasmic space or in the cytoplasm of mutants rendered sufficiently oxidizing to allow disulphide-bonds to form, while proteins requiring post-translational modification such as glycosylation for stability or function have been expressed using the N-linked glycosylation system of *Campylobacter jejuni* engineered into *E. coli*.

Modified *E. coli* cells have been used in vaccine development, bioremediation, production of biofuels, lighting, and production of immobilised enzymes.

K-12 is a mutant form of E-coli that over-expresses the enzyme Alkaline Phosphatase (ALP). The mutation arises due to a defect in the gene that constantly codes for the enzyme. A gene that is producing a product without any inhibition is said to have constitutive activity. This particular mutant form is used to isolate and purify the aforementioned enzyme.

Model Organism

E. coli is frequently used as a model organism in microbiology studies. Cultivated strains (e.g. *E. coli* K12) are well-adapted to the laboratory environment, and, unlike wild-type strains, have lost their ability to thrive in the intestine. Many laboratory strains lose their ability to form biofilms. These features protect wild-type strains from antibodies and other chemical attacks, but require a large expenditure of energy and material resources.

In 1946, Joshua Lederberg and Edward Tatum first described the phenomenon known as bacterial conjugation using *E. coli* as a model bacterium, and it remains the primary model to study conjugation.*E. coli* was an integral part of the first experiments to understand phage genetics, and early researchers, such as Seymour Benzer, used *E. coli* and phage T4 to understand the topography of gene structure. Prior to Benzer's research, it was not known whether the gene was a linear structure, or if it had a branching pattern.

E. coli was one of the first organisms to have its genome sequenced; the complete genome of *E. coli* K12 was published by *Science* in 1997.

By evaluating the possible combination of nanotechnologies with landscape ecology, complex habitat landscapes can be generated with details at the nanoscale.On such synthetic ecosystems, evolutionary experiments with *E. coli* have been performed to study the spatial biophysics of adaptation in an island biogeography on-chip.

Studies are also being performed attempting to program *E. coli* to solve complicated mathematics problems, such as the Hamiltonian path problem.

History

In 1885, the German-Austrian pediatrician Theodor Escherich discovered this organism in the feces of healthy individuals. He called it *Bacterium coli commune* because it is found in the colon. Early classifications of prokaryotes placed these in a handful of genera based on their shape and motility (at that time Ernst Haeckel's classification of bacteria in the kingdom Monera was in place).

*Bacterium coli*was the type species of the now invalid genus *Bacterium* when it was revealed that the former type species ("*Bacterium triloculare*") was missing. Following a revision of *Bacterium*, it was reclassified as *Bacillus coli* by Migula in 1895 and later reclassified in the newly created genus *Escherichia*, named after its original discoverer.

Wheat Germ Extract (WGE)

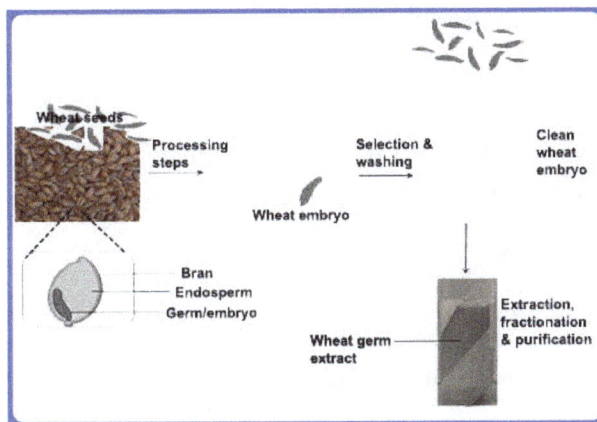

Wheat germ extract (WGE): This is a cell-free expression system that is capable of producing full-length proteins with correct folding and PTMs from bacterial, plant or animal sources. Yields obtained in this system are however slightly lower than the E. coli and RRL.

The wheat germ extract is also used for in vitro translation. The extract is from plant embryo and it contains all the machinery for protein synthesis such as initiation factors, translation factors etc. In addition to having all the components needed for protein expression, wheat germ extract has very low levels of endogenous mRNA, which aid in reducing the background translation by a considerable amount. WGE has been used for efficient translation of exogenous RNA from a variety of different organisms.

Rabbit Reticulocyte Lysate (RRL)

Another commonly used system for in vitro translation of proteins is the rabbit reticulocyte lysate. In vivo, the reticulocytes have the entire translation machinery as they produce hemoglobin. Thus, the reticulocyte lysate is ideal for translation of proteins, as it contains all the requirements for protein expression – initiation factors ribosomes etc. Addition of micrococcal nuclease to the extract results in degradation of the endogenous mRNA.

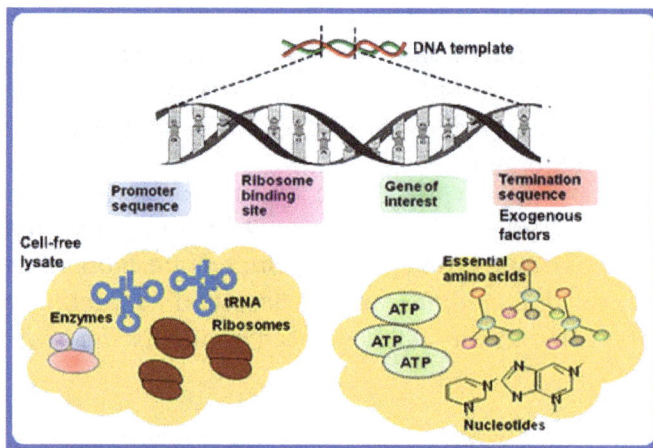

Rabbit reticulocyte lysate (RRL) is a mammalian cell-free system, which is more suitable for expression of full-length eukaryotic proteins that require proper folding and PTMs.

Applications

The microarrays spotted with proteins expressed using cell-free expression system have variety of applications such as:

- Biomarker discovery

- Immunogenicity studies

- Protein-protein interaction studies

- Post-translational modification studies

- Simultaneous screening of a large number of proteins

Advantages and Challenges

Technique	Advantages	Challenges
PISA	• Creates soluble proteins *in situ*. • Overcomes common problems (Protein insolubility, degradation and aggregation issues) endemic during protein expression in prokaryotic systems. • Only tagged protein remains on the array.	• Requires immediate utilization of PCR-produced DNA. • Not cost effective as a relatively large volume of cell-free lysate is required. • Hexa-histidine tag may interfere with proper protein folding.
NAPPA	• Use of mammalian expression systems allows efficient folding. • Access to a wide variety of cloned cDNAs. • Shelf-life not an issue: cDNA arrays stable for long periods at RT, and more difficult-to-store proteins produced just before assay. • Cost effective: low required volume of cell free lysate. • Effective process: over 95% of proteins tested express and capture well. Discrete spots are obtained.	• Need for time-consuming cloning before generating array, or alternately dependent on available clones. • Need to clone gene of interest as its GST fusion • Pure protein arrays not produced: expressed protein remains co-localized to cDNA. • Peptide tags may produce steric hindrance while studying protein interactions. • Correct functionality of proteins always remains in doubt during cell-free expression.
MIST	• Non-purified PCR product can be used as DNA template. • Very high-density protein arrays generated.	• Non-specific protein binding can occur. • Time consuming. • Loss of signal intensity occurs with prolonged incubation.
DAPA	• A pure protein array, free of DNA is generated. • Allows the generation of multiple protein arrays from a single DNA template. • The DNA template array can be stored for long periods.	• Broadening of spots may occur due to protein diffusion. • Multimeric proteins may not assemble efficiently. • Time consuming process.
Halo-link Protein arrays	• Covalent bond allows firm immobilization of proteins. • Proteins are captured in an oriented manner with no non-specific adsorption. • Little functional or quantitative losses of materials during washing steps. • Accurate quantification of protein possible. • No requirement for microarray printer as gaskets for printing are provided.	• Has not been validated for high density-large protein number arrays, although theoretically possible. • Loss of protein function may occur due to binding to the Halotag.

The usage of protein microarrays have made it possible to screen a large number of proteins simultaneously, leading to high-throughput studies. Many types of cell-free expression based microarrays have shown promising results.

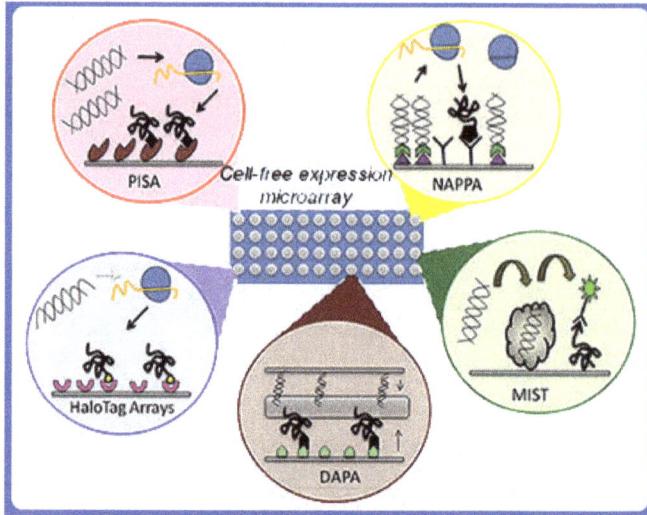

An overview of cell-free expression based microarrays.

NAPPA and DAPA aid in building pure protein arrays with no DNA contamination. Proteins have relatively shorter shelf-lives and protein microarrays help in increasing the shelf-life of the proteins. Making protein arrays is also cost-effective, as very small amounts of reagents are used. Moreover, the DNA templates are reusable, which reduces the cost further. The production of proteins becomes much easier, and storing of DNA arrays helps in reproducing the arrays when required. Microarrays have tremendous potential in clinical applications as well as non-clinical research such as detection of protein-protein interaction and biological screening.

Nucleic Acid Programmable Protein Arrays

Protein microarrays consist of solid surfaces on which the proteins of interest are spotted. Earlier, each protein used for spotting on the slide surface using chemical techniques. This required manual production of each protein, which was very tedious. The self assembling protein microarray or NAPPA approach introduced by Ramachandran et al. (2004) used spotting of expression plasmids containing cDNAs of interest on the array surface and expression of proteins in situ by a mammalian cell free expression system at the time of assay. All proteins were expressed with fusion tags, which correspond to capture agents printed along with the plasmid DNA and used to capture the protein immediately after translation. By producing the proteins just-in-time for assay the opportunity for its denaturation is significantly reduced, and the use of mammalian transcription/translation system encouraged natural protein folding for mammalian protein. The NAPPA microarray is a highly innovative cell-free expression based technology, which helps in production of thousands of proteins simultaneously, and also capturing them on the slide to form the array as and when they are formed.

In NAPPA, cDNA (containing a tag, usually GST) of the desired proteins are spotted along-with BSA, BS3 and capture antibody (anti-GST antibody) on functionalized slide. For activation of the array, a cell-free expression mixture containing in vitro transcription and translation mix (rabbit reticulocyte lysate, T7 polymerase, amino acid mixtures, RNase inhibitor etc.) is added onto the slide. This leads to the production of desired proteins from the cDNA. The protein produced contains a tag, which binds to the capture agent coated on the slide. Thus, a protein replica is formed in place of the DNA array.

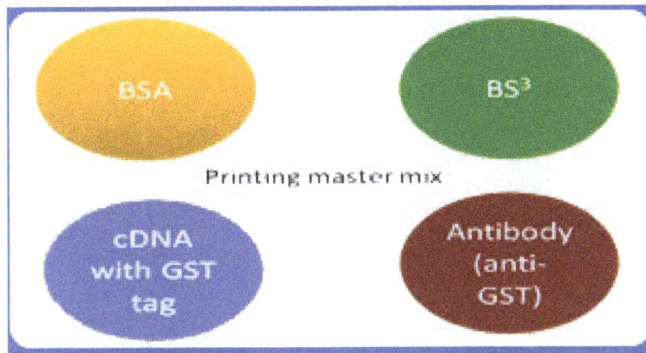

Master mix for NAPPA printing consists of cDNA with GST tag, BS3 cross linker, BSA and capture antibody (anti-GST).

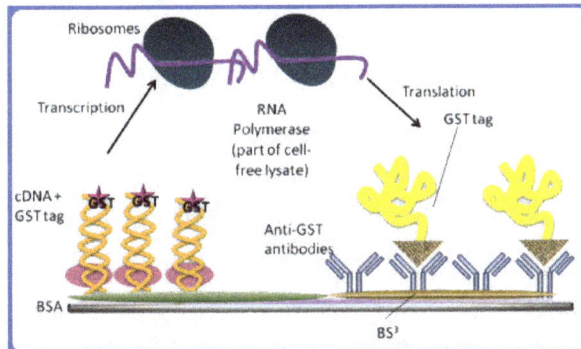

Working principle of Nucleic Acid Programmable Protein Arrays. The master-mix containing cDNA (with GST tag), BSA, BS3 and anti-GST antibody are printed on array surface. Adding RRL to the arrays carries out protein expression. The newly synthesized proteins are captured by antibody through the GST tag and protein microarray is produced.

Workflow of Nappa

To ensure construction of NAPPA protein microarray, one requires careful design of each of the components. The design and role of each of the components of NAPPA workflow are as follows:

Preparation of Master-mix for Printing

The solid support for printing the array is generally a glass slide. But other supports such as gold, nitrocellulose, hydrogel also may be used. Depending upon the surface of the microarray, the surface chemistry for printing cDNA differs.

- The glass slide is coated with APTES (aminopropyltriethoxysilane or aminosilane) and it contains a large amount of positively charged amino groups, which bind to the negatively charged phosphate groups on DNA. The slides are exposed to UV or baked at 85°C enabling the strong covalent attachment of DNA on the silane coated slides.

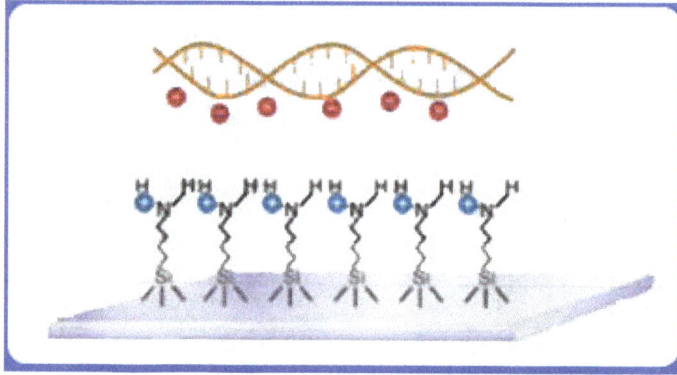

Aminosilane coating of the glass slide. The positive groups of amine on the slide surface covalently bind to the negatively charged phosphate groups of DNA, which enables immobilization of DNA molecules onto the slide surface.

- If the cDNA is biotinylated (using psoralen-biotin), the slide is coated with avidin along with aminosilane. This leads to a strong biotin-avidin binding and this interaction enables the DNA immobilization onto the slide surface.

- For protein capture, a BS3 cross-linker and anti-tag antibody is also spotted on the silane coated glass slide. This linker helps in formation of amine-amine bonds, and thus is used for tethering the capture antibody on the glass slide.

- The cDNA for the protein of interest is designed for two purposes; 1) the cDNA should be captured on the slide before protein production, and 2) it must have a tag, which helps in identification and capture of the expressed protein. The cDNA plasmids are designed to form a fusion tag at either the N or C terminal. Generally, plasmids are designed to add a C-terminal glutathione S-transferase (GST) tag.

- Bovine Serum Albumin (BSA) is added to this master mix as it enhances the binding of the DNA to the slide.

- The master-mix is printed on the slide by manual spotting or using an automated microarrayer.

Protein Production and Capture

For protein production using in vitro transcription and translation system (IVTT), a cell-free expression mixture containing rabbit reticulocyte lysate, T7 polymerase, amino acid mixtures and RNase inhibitor is added onto the slide. The proteins expressed from the cDNA contain a tag (usually GST tag) at the C-terminal of the protein. The anti-GST antibody is coated on the aminosilane slide, with the BS3 cross-linker. The synthesized protein gets immobilized to the slide by antigen-antibody reactions and protein array is formed.

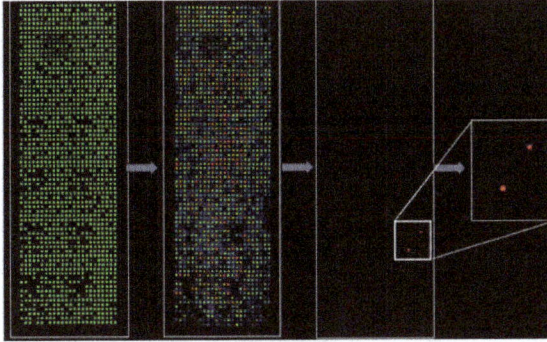

Quality control of NAPPA array: PicoGreen dye is used for testing DNA printing quality, Ani-GST antibody is used for testing protein expression. The protein specific antibody is used to test protein-specific expression.

Applications of Nappa

Autoantibody Biomarker Screening

Antibodies to tumor antigens have advantages over other serum proteins as potential cancer biomarkers as they are stable, highly specific, easily purified from serum, and are readily detected with well-validated secondary reagents. The antibodies directed at self-antigens are referred to as autoantibodies. NAPPA arrays have been used to identify autoantibody biomarkers in sera that can be readily used for the early detection of cancers. For autoantibody screening, samples such as serum or cell lysate is added on the chip. If target antigens are present in the sample, autoantibodies bind to their targets, and this can be detected using a labeled anti-IgG antibody.

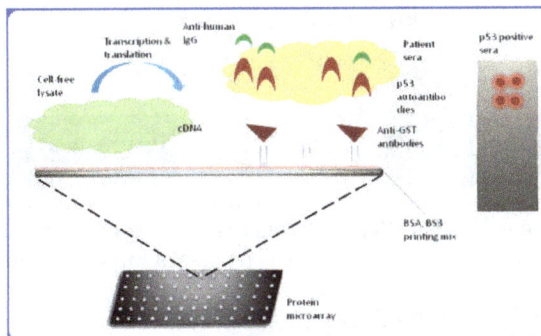

Detection of p53 autoantibodies using NAPPA microarrays

Protein Interactions - One key application of NAPPA is to test protein-protein interactions. Typically, this is done by probing an array of proteins with a purified query protein. NAPPA is also able to employ co-expression of the target and the query protein by transcribing and translating them in the same extract. To do this, simply appropriate query DNA was added into the IVTT before applying it to the array. For fluorescent detection, the query proteins were expressed with an epitope tag that is different from the one used to capture the target proteins on the array. Following the co-expression, the binding of the query to a specific target protein was detected by using an antibody to the query epitope.

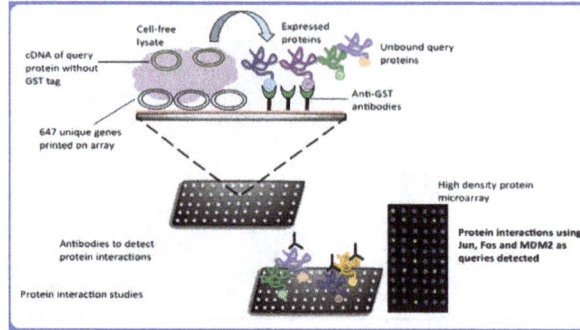

Protein-protein interaction study using NAPPA arrays

Advantages and Challenges

NAPPA is a promising cell-free expression based protein microarray technology. This technique uses mammalian expression systems and thus allows efficient protein folding. Access to a wide variety of cloned cDNAs, allows spotting of almost any protein on the array. One of the most prominent advantages of NAPPA is long shelf-life of the arrays. NAPPA arrays are very cost-effective as the volume of the cell-free lysate and the template is very low. NAPPA is very effective process as 95% of the proteins are usually expressed and captured on the slide.

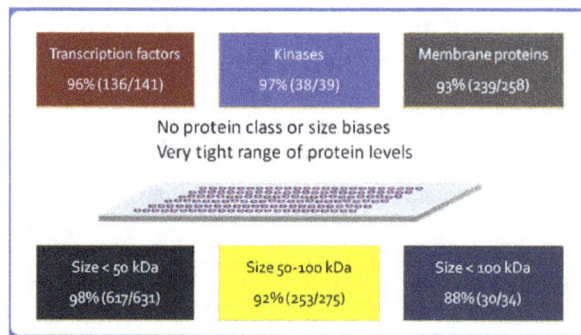

Success of protein expression using NAPPA protein microarrays. Protein expression in kinases, transcription factor and membrane proteins (>95%). There is no bias of low or high molecular weight protein

Nappa Approach Offers Following Advantages Over Traditional Methods

- Replaces printing proteins with the more reliable process of printing DNA

- Avoids the need to express and purify proteins

- Avoids concerns about protein shelf life because the proteins are made fresh at the time of assay

- Protein integrity: uses mammalian machinery to synthesize proteins

- No tedious gene introduction into mammalian cells

- Interaction is not limited to context of nucleus

- Gene toxicity and auto activation of reporter genes are not an issue

- Expensive mass spectrometers are not required

- Analysis of multimeric complexes

- Post translational modifications

- Protein production and obtaining data in real time

- Sequence-verified plasmid templates provide good quality controls

- Attachment/detection scheme changeable

However, like any other technology, NAPPA has certain limitations. The process is time consuming, as cloning of each cDNA template needs to be done before generating array, or alternately it depends on availability of clones. Each of the cDNA needs to be cloned with fusion tags (e.g. GST). Another problem associated with NAPPA is that the production of pure protein arrays is not possible, as the expressed protein remains co-localized with cDNA. The peptide tags may produce steric hindrance while studying protein interactions. Finally, It is difficult to assess the functionality of expressed proteins after in vitro transcription and translation steps.

NAPPA is very effective technique for production of protein microarrays. It has already been used in various studies for immunological screening, screening of biomarkers and to study protein-protein interaction. The arrays can be stored for long, as NAPPA uses DNA for printing, and proteins can be produced on demand. Integration of NAPPA microarrays with label-free detection techniques such as surface Plasmon resonance imaging would be interesting future direction for several high throughput proteomic applications.

Microarrays for PTM Analysis

Significance of Post-Translational Modifications

Proteome is typically of higher complexity than genome. Although, the human genome comprised of ~ 25,000 genes, the total number of proteins in the human proteome is comparatively very large. This indicates that more than one protein can be expressed by a single gene (IHGSC 2004; Jensen et al 2004). Different mRNA transcripts are generated through genomic recombination, using different transcription initiation and/or termination site, and by alternative splicing from a single gene (Ayoubi et al. 1996). Another level of complexity in the human proteome is generated by PTMs. PTMs are chemical modifications, which occur at side chains of amino acids or at peptide linkage of a protein by enzymatic activity, and these modifications play a key role in functional proteomics as they regulate cellular activities, localization and interaction of proteins with other cellular molecules. It is estimated that 5% of the genes in the human genome encode enzymes that perform more than 300 types of post-translational modifications (Walsh et al. 2006).

The chemical modifications that take place at certain amino acid residues after the protein is synthesized by translation are known as post-translational modifications (PTMs). These are essential for normal functioning of the protein. There are several types of post-translational modifications that can take place at different amino acid residues. The most commonly observed PTMs include phosphorylation, glycosylation, methylation as well as hydroxylation and acylation. Many of these modifications, particularly phosphorylation, serve as regulatory mechanisms for protein action.

PTM generate tremendous diversity and are extremely important. Many documented effects of PTMs include – change in enzymatic activity, ability to interact with other proteins, sub-cellular localization and targeted degradation. Phosphorylation of amino acid residues is carried out by a class of enzymes known as kinases that most commonly modify side chains of amino acids containing a hydroxyl group. Phosphorylation requires presence of a phosphate donor molecule such as ATP, GTP or other phoshorylated substrates. Removal of phosphate groups are carried out by phosphatase enzyme and it forms an important mechanism for regulation of proteins. Glycosylation involves linking saccharides to proteins in presence of glycosyltransferases enzymes, giving rise to a glycoprotein. Glycosylation plays vital role in various biological functions such as antigenicty of immunological molecules, cell division, protein targeting stability and interactions. Aberrant glycosylation forms result into various human congenital disorders. Some of the most important PTMs are listed below.

Type of post-translational modification and significance

Type of post-translational modification	Amino acid residue involved	Significance
Phosphorylation and dephosphorylation	Serine, Threonine, and Tyrosine	Regulate cellular processes including cell cycle, growth, apoptosis and signal transduction pathways.
Glycosylation	Aspargine, serine, Threonine	Influence protein folding, conformation, distribution, stability and activity
Ubiquitination	Lysine	Mediate degradation of protein
Methylation	Nitrogen or oxygen to amino acid side chains (lysine)	Epigenetic regulation N-methylation is irreversible while O-methylation is potentially reversible
S-Nitrosylation	Cysteine	Stabilize proteins, regulate gene expression and provide NO donors

N-Acetylation	N-terminal methi-onine	Biological significance of N-acetylation is not clear
	μ-NH2 of lysine	Regulation of Gene expression
N-myristoylation	N-terminal glycine	Target proteins to plasma membranes
S-palmitoylation	Thiolate side chain of cysteine	Target proteins to plasma membranes
S-prenylation	Cysteine residues within 5 amino acids from the C-terminus	Proteolytic cleavage by Rce1 and methylation by isoprenyl cysteine methyl transferase (ICMT)

Methods for Studying Post-translational Modifications

Studying post-translational modifications is particularly important as PTMs plays pivotal role in cellular physiology and gets altered in diseases. Few of the PTMs have been extensively investigated at the proteome level. The challenges in studying PTMs are the development of specific detection and purification methods. With continuous advancement in technologies, it is possible to meet these challenges. Various methods can be employed for studying the PTMs, such as mass spectrometry (MS) based proteomics technologies for global PTM analysis, Western blotting and protein microarrays. Antibody microarray provides a useful platform for PTM analysis in a high-throughput manner. There are various types of microarrays specialized for the identification of specific PTMs.

Microarray Platform for PTM Analysis

Type of Microarray	Applications
Peptide array	Characterization of kinase specificity
Antibody arrays	Network analysis of phosphorylated proteins
Oligosaccharide microarrays	Protein–carbohydrates interactions
Lectin arrays	Characterizing glycoproteins

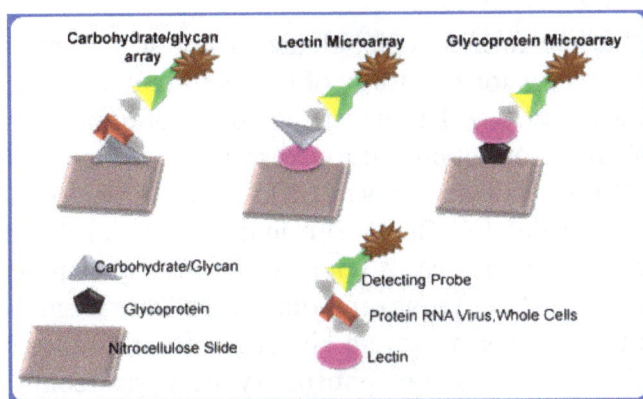

For glycosylation studies different types of microarray platforms such as carbohydrate array, lectin arrays, glycoprotein arrays and other array formats have been used.

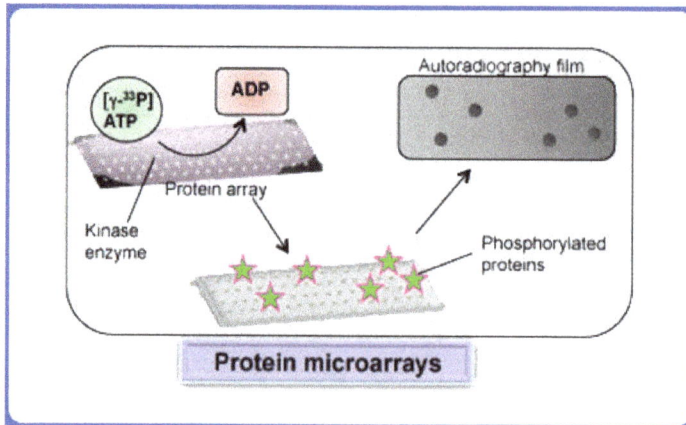

Protein microarrays for PTM analysis

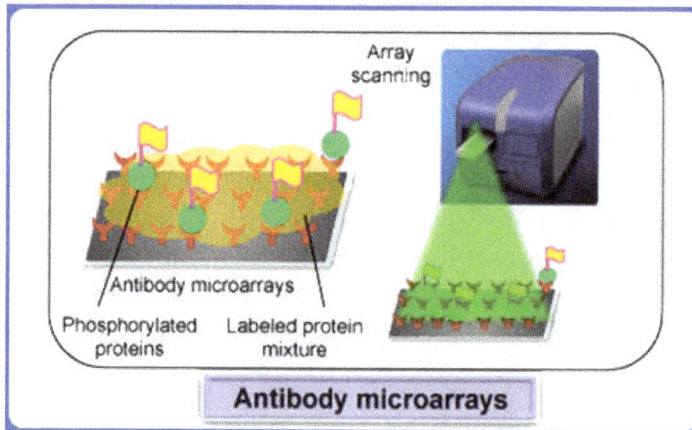

Antibody microarrays for PTM analysis

Applications of Microarrays in Studying Post-Translational Modifications

Global analysis of protein phosphorylation in Yeast

Snyder and colleagues performed a global analysis of protein phosphorylation, using whole proteome arrays for the study of protein kinases in Saccharomyces cerevisiae (Ptacek, J. et al., 2005). There were 5800 out of 6200 open reading frames (ORFs) from yeast that were cloned into expression vector carrying two tags; an N-terminus gluthathione S-transferase (GST) and oligohistidine tags. Proteins were expressed and purified by affinity chromatography. Proteins were spotted in duplicate onto nickel-coated slides and purified kinases were used to modify them. Phosphorylation substrates on yeast proteome chips were identified by incubating slides with individually 87 yeast protein kinases in the presence of γ-33P ATP. To identify proteins those undergo auto-phosphorylation, proteome chips were incubated with γ-33P ATP in kinase buffer and phosphorylation was detected using auto-radiography.

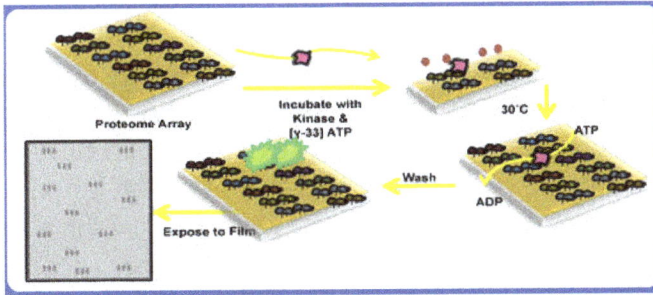

Schematic of PTM analysis to identify kinase substrates: yeast proteome array was prepared by spotting about 4,400 proteins in duplicates and each kinase was over-expressed, purified and assayed on yeast proteome chips and detection was done autoradiographically.

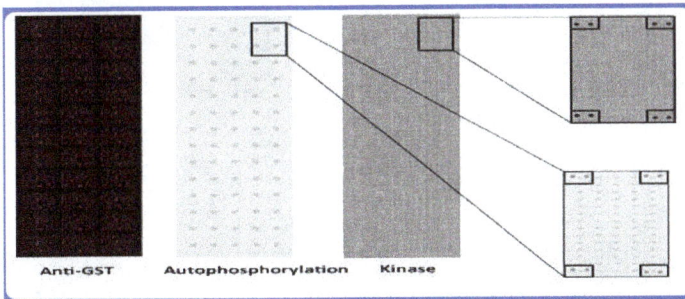

Kinase assays on protein chips; slide on the left side represents the quality of proteome chip as all fusion proteins has GST tag; middle slide shows the autophosphorylation signals and four corners of the two boxes of the slides on the right side show signal for auto-phosphorylation from kinases printed on slides.

This study created a map of yeast phosphoproteome and identified several phosphorylation events. Protein kinases showed specificity for substrates and 73% of substrates were recognized by specific few kinases, indicating a strong preference of kinases for specific substrates.

Detection of lysine methylation on proteins using microarrays

Lysine methylation is one of the key post-translational modifications, which plays an important role in several biological activities, including epigenetic regulation. Levy et al. (2011) used the ProtoArray platform, which contained more than 9500 human proteins and identified new substrates for protein lysine methyltransferase (PKMT) enzymes, SETD6 and SETD7. SETD6 methylates Rel A on lys 310 residue, and SETD7 is a histone methyltransferase, which methylates histone H3 at lys4 position. Fluorescent (Pan-methyl antibody) and radioactive assays (radioactively labeled SAM) were used for detection. Fluorescent-based detection identified 321 positive candidates for SETD7, while 118 positive candidates for SETD6 and results were consistent with radioactivity based method. 26 candidates were found to be common for SETD6 in both the detection methods. Further, in vitro validation was performed for six candidates (TCEA1, PAK4, RPS27L, SFRS2, PLK1, and DNAJC8) and in vivo validation for PLK1 and PAK4 in 293T cells by immuno-precipitation and western blotting.

Identification of protein kinase substrates using protein microarray

Protein arrays have been successfully used for identification of substrate for kinases. In one such study, Meng et al (2008) identified 23 substrates and further validated CAMK2, while 4 candidates (FLJ22795, SH3YL1, CRKL, and ABI1) were identified as substrate for ABL kinase. It was demonstrated that time, buffer compositions, and protein concentrations affect the assay; therefore assay parameters should be optimized before carrying out functional assay.

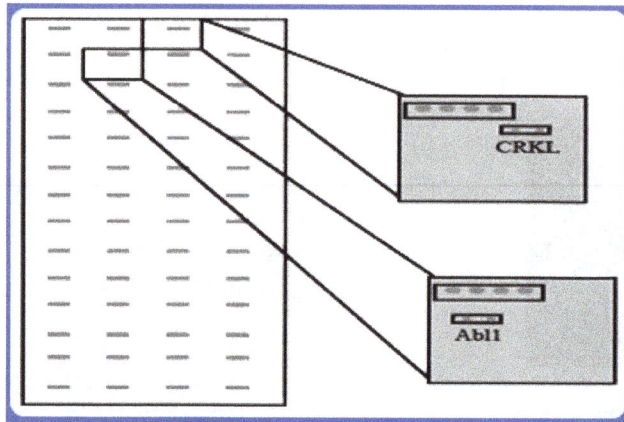

Identification of Abl kinase substrate using protein microarrays

In 2009, Yifat Merbl and Marc W. Kirschner (2009) have successfully used microarray approach to study substrates for ubiquitination using cell extracts prepared from cells released from mitotic checkpoint.

Challenges

PTMs regulate catalytic activities of several proteins thus influence their biological functions. The protein microarray platforms have shown promising results for PTM study; however, these approaches have certain limitations:

- Limited number of proteins due to difficulty in protein expression and purification

- Tethering of proteins on surface may results in loss of their functionality

- Unavailability of specific antibodies

- Sensitivity and dynamic range of detection systems

- Complications in studying PTMs (other than phosphorylation) such as ubiquitination, methylation etc which play important role in biological activities

- Microarray platform cannot portray in vivo conditions hence findings should be validated by other confirmatory experiments

Phage Display

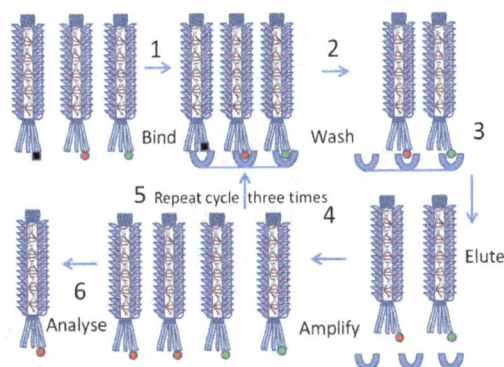

The sequence of events that are followed in phage display screening to identify polypeptides that bind with high affinity to desired target protein or DNA sequence.

Phage display is a laboratory technique for the study of protein–protein, protein–peptide, and protein–DNA interactions that uses bacteriophages (viruses that infect bacteria) to connect proteins with the genetic information that encodes them. In this technique, a gene encoding a protein of interest is inserted into a phage coat protein gene, causing the phage to "display" the protein on its outside while containing the gene for the protein on its inside, resulting in a connection between genotype and phenotype. These displaying phages can then be screened against other proteins, peptides or DNA sequences, in order to detect interaction between the displayed protein and those other molecules. In this way, large libraries of proteins can be screened and amplified in a process called *in vitro* selection, which is analogous to natural selection.

The most common bacteriophages used in phage display are M13 and fd filamentous phage, though T4,T7, and λ phage have also been used.

History

Phage display was first described by George P. Smith in 1985, when he demonstrated the display of peptides on filamentous phage by fusing the peptide of interest onto gene III of filamentous phage. A patent by George Pieczenik claiming priority from 1985 also describes the generation of phage display libraries. This technology was further developed and improved by groups at the Laboratory of Molecular Biology with Greg Winter and John McCafferty, The Scripps Research Institute with Lerner and Barbas and the German Cancer Research Center with Breitling and Dübel for display of proteins such as antibodies for therapeuticprotein engineering.

Principle

Like the two-hybrid system, phage display is used for the high-throughput screening of protein interactions. In the case of M13 filamentous phage display, the DNA encoding the protein or peptide of interest is ligated into the pIII or pVIII gene, encoding either the minor or major coat protein, respectively. Multiple cloning sites are sometimes used to ensure that the fragments are inserted in all three possible reading frames so that the cDNA fragment is translated in the proper frame. The phage gene and insert DNA hybrid is then inserted (a process known as "transduction") into *Escherichia coli* (E. coli) bacterial cells such as TG1, SS320, ER2738, or XL1-Blue *E. coli*. If a "phagemid" vector is used (a simplified display construct vector) phage particles will not be released from the *E. coli* cells until they are infected with helper phage, which enables packaging of the phage DNA and assembly of the mature virions with the relevant protein fragment as part of their outer coat on either the minor (pIII) or major (pVIII) coat protein. By immobilizing a relevant DNA or protein target(s) to the surface of a microtiter plate well, a phage that displays a protein that binds to one of those targets on its surface will remain while others are removed by washing. Those that remain can be eluted, used to produce more phage (by bacterial infection with helper phage) and so produce a phage mixture that is enriched with relevant (i.e. binding) phage. The repeated cycling of these steps is referred to as 'panning', in reference to the enrichment of a sample of gold by removing undesirable materials. Phage eluted in the final step can be used to infect a suitable bacterial host, from which the phagemids can be collected and the relevant DNA sequence excised and sequenced to identify the relevant, interacting proteins or protein fragments.

The use of a helper phage can be eliminated by using 'bacterial packaging cell line' technology.

Elution can be done combining low-pH elution buffer with sonification, which, in addition to loosening the peptide-target interaction, also serves to detach the target molecule from the immobilization surface. This ultrasound-based method enables single-step selection of a high-affinity peptide.

Applications

Applications of phage display technology include determination of interaction partners of a protein (which would be used as the immobilised phage "bait" with a DNA library consisting of all coding sequences of a cell, tissue or organism) so that the function or the mechanism of the function of that protein may be determined. Phage display is also a widely used method for *in vitro* protein evolution (also called protein engineering). As such, phage display is a useful tool in drug discovery. It is used for finding new ligands (enzyme inhibitors, receptor agonists and antagonists) to target proteins. The technique is also used to determine tumour antigens (for use in diagnosis and therapeutic targeting) and in searching for protein-DNA interactions using specially-constructed DNA libraries with randomised segments.

Competing methods for *in vitro* protein evolution include yeast display, bacterial display, ribosome display, and mRNA display.

Antibody Maturation *in Vitro*

The invention of antibody phage display revolutionised antibody drug discovery. Initial work was done by laboratories at the MRC Laboratory of Molecular Biology (Greg Winter and John McCafferty), the Scripps Research Institute (Richard Lerner and Carlos F. Barbas) and the German Cancer Research Centre (Frank Breitling and Stefan Dübel). In 1991, The Scripps group reported the first display and selection of human antibodies on phage. This initial study described the rapid isolation of human antibody Fab that bound tetanus toxin and the method was then extended to rapidly clone human anti-HIV-1 antibodies for vaccine design and therapy.

Phage display of antibody libraries has become a powerful method for both studying the immune response as well as a method to rapidly select and evolve human antibodies for therapy. Antibody phage display was later used by Carlos F. Barbas at The Scripps Research Institute to create synthetic human antibody libraries, a principle first patented in 1990 by Breitling and coworkers (Patent CA 2035384), thereby allowing human antibodies to be created in vitro from synthetic diversity elements.

Antibody libraries displaying millions of different antibodies on phage are often used in the pharmaceutical industry to isolate highly specific therapeutic antibody leads, for development into antibody drugs primarily as anti-cancer or anti-inflammatory therapeutics. One of the most successful was HUMIRA (adalimumab), discovered by Cambridge Antibody Technology as D2E7 and developed and marketed by Abbott Laboratories. HUMIRA, an antibody to TNF alpha, was the world's first fully human antibody, which achieved annual sales exceeding $1bn.

General Protocol

Below is the sequence of events that are followed in phage display screening to identify polypeptides that bind with high affinity to desired target protein or DNA sequence:

1. Target proteins or DNA sequences are immobilised to the wells of a microtiter plate.

2. Many genetic sequences are expressed in a bacteriophage library in the form of fusions with the bacteriophage coat protein, so that they are displayed on the surface of the viral particle. The protein displayed corresponds to the genetic sequence within the phage.

3. This phage-display library is added to the dish and after allowing the phage time to bind, the dish is washed.

4. Phage-displaying proteins that interact with the target molecules remain attached to the dish, while all others are washed away.

5. Attached phage may be eluted and used to create more phage by infection of suitable bacterial hosts. The new phage constitutes an enriched mixture, containing considerably less irrelevant phage (i.e. non-binding) than were present in the initial mixture.

6. Steps 3 to 6 are optionally repeated one or more times, further enriching the phage library in binding proteins.

7. Following further bacterial-based amplification, the DNA within in the interacting phage is sequenced to identify the interacting proteins or protein fragments.

Selection of the Coat Protein

Filamentous Phages

pIII

pIII is the protein that determines the infectivity of the virion. pIII is composed of three domains (N1, N2 and CT) connected by glycine-rich linkers. The N2 domain binds to the F pilus during virion infection freeing the N1 domain which then interacts with a TolA protein on the surface of the bacterium. Insertions within this protein are usually added in position 249 (within a linker region between CT and N2), position 198 (within the N2 domain) and at the N-terminus (inserted between the N-terminal secretion sequence and the N-terminus of pIII). However, when using the BamHI site located at position 198 one must be careful of the unpaired Cysteine residue (C201) that could cause problems during phage display if one is using a non-truncated version of pIII.

An advantage of using pIII rather than pVIII is that pIII allows for monovalent display when using a phagemid (Ff-phage derived plasmid) combined with a helper phage. Moreover, pIII allows for the insertion of larger protein sequences (>100 amino acids) and is more tolerant to it than pVIII. However, using pIII as the fusion partner can lead to a decrease in phage infectivity leading to problems such as selection bias caused by difference in phage growth rate or even worse, the phage's inability to infect its host. Loss of phage infectivity can be avoided by using a phagemid plasmid and a helper phage so that the resultant phage contains both wild type and fusion pIII.

cDNA has also been analyzed using pIII via a two complementary leucine zippers system, Direct Interaction Rescue or by adding an 8-10 amino acid linker between the cDNA and pIII at the C-terminus.

pVIII

pVIII is the main coat protein of Ff phages. Peptides are usually fused to the N-terminus of pVIII. Usually peptides that can be fused to pVIII are 6-8 amino acids long. The size restriction seems to have less to do with structural impediment caused by the added section and more to do with the size exclusion caused by pIV during coat protein export. Since there are around 2700 copies of the protein on a typical phages, it is more likely that the protein of interest will be expressed polyvalently even if a phagemid is used. This makes the use of this protein unfavorable for the discovery of high affinity binding partners.

To overcome the size problem of pVIII, artificial coat proteins have been designed. An example is Weiss and Sidhu's inverted artificial coat protein (ACP) which allows the display of large proteins at the C-terminus. The ACP's could display a protein of 20kDa, however, only at low levels (mostly only monovalently).

pVI

pVI has been widely used for the display of cDNA libraries. The display of cDNA libraries via phage display is an attractive alternative to the yeast-2-hybrid method for the discovery of interacting proteins and peptides due to its high throughput capability. pVI has been used preferentially to pVIII and pIII for the expression of cDNA libraries because one can add the protein of interest to the C-terminus of pVI without greatly affecting pVI's role in phage assembly. This means that the stop codon in the cDNA is no longer an issue. However, phage display of cDNA is always limited by the inability of most prokaryotes in producing post-translational modifications present in eukaryotic cells or by the misfolding of multi-domain proteins.

While pVI has been useful for the analysis of cDNA libraries, pIII and pVIII remain the most utilized coat proteins for phage display.

pVII and pIX

In an experiment in 1995, display of Glutathione S-transferase was attempted on both pVII and pIX and failed. However, phage display of this protein was completed successfully after the addition of a periplasmic signal sequence (pelB or ompA) on the N-terminus. In a recent study, it has been shown that AviTag, FLAG and His could be displayed on pVII without the need of a signal sequence. Then the expression of single chain Fv's (scFv), and single chain T cell receptors (scTCR) were expressed both with and without the signal sequence.

PelB (an amino acid signal sequence that targets the protein to the periplasm where a signal peptidase then cleaves off PelB) improved the phage display level when compared to pVII and pIX fusions without the signal sequence. However, this led to the incorporation of more helper phage genomes rather than phagemid genomes. In all cases, phage display levels were lower than using pIII fusion. However, lower display might be more favorable for the selection of binders due to lower display being closer to true monovalent display. In five out of six occasions, pVII and pIX fusions without pelB was more efficient than pIII fusions in affinity selection assays. The paper even goes on to state that pVII and pIX display platforms may outperform pIII in the long run.

The use of pVII and pIX instead of pIII might also be an advantage because virion rescue may be undertaken without breaking the virion-antigen bond if the pIII used is wild type. Instead, one could cleave in a section between the bead and the antigen to elute. Since the pIII is intact it does not matter whether the antigen remains bound to the phage.

T7 Phages

The issue of using Ff phages for phage display is that they require the protein of interest to be translocated across the bacterial inner membrane before they are assembled into the phage. Some proteins cannot undergo this process and therefore cannot be displayed on the surface of Ff phages. In these cases, T7 phage display is used instead. In T7 phage display, the protein to be displayed is attached to the C-terminus of the gene 10 capsid protein of T7.

The disadvantage of using T7 is that the size of the protein that can be expressed on the surface is limited to shorter peptides because large changes to the T7 genome cannot be accommodated like it is in M13 where the phage just makes its coat longer to fit the larger genome within it. However, it can be useful for the production of a large protein library for scFV selection where the scFV is expressed on an M13 phage and the antigens are expressed on the surface of the T7 phage.

Bioinformatics Resources and Tools

Databases and computational tools for mimotopes have been an important part of phage display study. Databases, programs and web servers have been widely used to exclude target-unrelated peptides, characterize small molecules-protein interactions and map protein-protein interactions. Users can use three dimensional structure of a protein and the peptides selected from phage display experiment to map conformational eptiopes. Some of the fast and efficient computational methods are available online.

Bacteriophage

A bacteriophage is a virus that infects and replicates within a bacterium. The term is derived from "bacteria". Bacteriophages are composed of proteins that encapsulate a DNA or RNA genome, and may have relatively simple or elaborate structures. Their genomes may encode as few as four genes, and as many as hundreds of genes. Phages replicate within the bacterium following the injection of their genome into its cytoplasm. Bacteriophages are among the most common and diverse entities in the biosphere. Bacteriophages are ubiquitous viruses, found wherever bacteria exist. It's estimated there are more than 10^{31} bacteriophages on the planet. That's ten million trillion trillion, more than every other organism on Earth, including bacteria, combined.

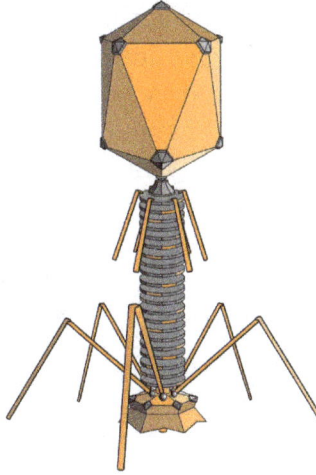

The structure of a typical myovirus bacteriophage

Phages are widely distributed in locations populated by bacterial hosts, such as soil or the intestines of animals. One of the densest natural sources for phages and other viruses is sea water, where up to 9×10^8 virions per milliliter have been found in microbial mats at the surface, and up to 70% of marine bacteria may be infected by phages. They have been used for over 90 years as an alternative to antibiotics in the former Soviet Union and Central Europe, as well as in France. They are seen as a possible therapy against multi-drug-resistant strains of many bacteria. Nevertheless, phages of Inoviridae have been shown to complicate biofilms involved in pneumonia and cystic fibrosis, and shelter the bacteria from drugs meant to eradicate disease and promote persistent infection.

Classification

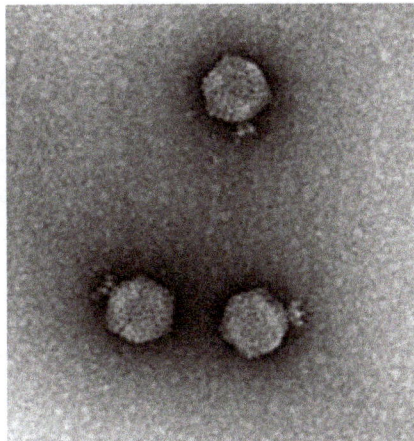

Bacteriophage P22, a member of the *Podoviridae* by morphology due to its short, non-contractile tail.

Bacteriophages occur abundantly in the biosphere, with different virions, genomes, and lifestyles. Phages are classified by the International Committee on Taxonomy of Viruses (ICTV) according to morphology and nucleic acid.

Nineteen families are currently recognized by the ICTV that infect bacteria and archaea. Of these, only two families have RNA genomes, and only five families are enveloped. Of the viral families with DNA genomes, only two have single-stranded genomes. Eight of the viral families with DNA genomes have circular genomes while nine have linear genomes. Nine families infect bacteria only, nine infect archaea only, and one (*Tectiviridae*) infects both bacteria and archaea.

ICTV classification of prokaryotic (bacterial and archaeal) viruses				
Order	**Family**	**Morphology**	**Nucleic acid**	**Examples**
Caudovi-rales	*Myoviridae*	Nonenveloped, contractile tail	Linear dsDNA	T4 phage, Mu, PBSX, P1Pu-na-like, P2, I3, Bcep 1, Bcep 43, Bcep 78
	Siphoviridae	Nonenveloped, noncon-tractile tail (long)	Linear dsDNA	λ phage, T5 phage, phi, C2, L5, HK97, N15
	Podoviridae	Nonenveloped, noncon-tractile tail (short)	Linear dsDNA	T7 phage, T3 phage, Φ29, P22, P37
Ligamenvi-rales	*Lipothrixviridae*	Enveloped, rod-shaped	Linear dsDNA	Acidianus filamentous virus 1
	Rudiviridae	Nonenveloped, rod-shaped	Linear dsDNA	Sulfolobus islandicus rod-shaped virus 1
Unassigned	*Ampullaviridae*	Enveloped, bottle-shaped	Linear dsDNA	
	Bicaudaviridae	Nonenveloped, lem-on-shaped	Circular dsD-NA	
	Clavaviridae	Nonenveloped, rod-shaped	Circular dsD-NA	
	Corticoviridae	Nonenveloped, isometric	Circular dsD-NA	
	Cystoviridae	Enveloped, spherical	Segmented dsRNA	
	Fuselloviridae	Nonenveloped, lem-on-shaped	Circular dsD-NA	
	Globuloviridae	Enveloped, isometric	Linear dsDNA	
	Guttaviridae	Nonenveloped, ovoid	Circular dsD-NA	
	Inoviridae	Nonenveloped, filamentous	Circular ssDNA	M13
	Leviviridae	Nonenveloped, isometric	Linear ssRNA	MS2, Qβ
	Microviridae	Nonenveloped, isometric	Circular ssDNA	ΦX174
	Plasmaviridae	Enveloped, pleomorphic	Circular dsD-NA	
	Tectiviridae	Nonenveloped, isometric	Linear dsDNA	

History

In 1896, Ernest Hanbury Hankin reported that something in the waters of the Ganges and Yamuna rivers in India had marked antibacterial action against cholera and could pass through a very fine porcelain filter. In 1915, BritishbacteriologistFrederick Twort, superintendent of the Brown Institution of London, discovered a small agent that infected and killed bacteria. He believed the agent must be one of the following:

1. a stage in the life cycle of the bacteria;

2. an enzyme produced by the bacteria themselves; or

3. a virus that grew on and destroyed the bacteria.

Twort's work was interrupted by the onset of World War I and shortage of funding. Independently, French-CanadianmicrobiologistFélix d'Hérelle, working at the Pasteur Institute in Paris, announced on 3 September 1917, that he had discovered "an invisible, antagonistic microbe of the dysentery bacillus". For d'Hérelle, there was no question as to the nature of his discovery: "In a flash I had understood: what caused my clear spots was in fact an invisible microbe ... a virus parasitic on bacteria." D'Hérelle called the virus a bacteriophage or bacteria-eater. He also recorded a dramatic account of a man suffering from dysentery who was restored to good health by the bacteriophages. It was D'Herelle who conducted much research into bacteriophages and introduced the concept of phage therapy.

In 1969, Max Delbrück, Alfred Hershey and Salvador Luria were awarded the Nobel Prize in Physiology and Medicine for their discoveries of the replication of viruses and their genetic structure.

Phage Therapy

Phages were discovered to be antibacterial agents and were used in the former Soviet Republic of Georgia (pioneered there by Giorgi Eliava with help from the co-discoverer of bacteriophages, Felix d'Herelle) and the United States during the 1920s and 1930s for treating bacterial infections. They had widespread use, including treatment of soldiers in the Red Army. However, they were abandoned for general use in the West for several reasons:

- Medical trials were carried out, but a basic lack of understanding of phages made these invalid.

- Antibiotics were discovered and marketed widely. They were easier to make, store and to prescribe.

- Former Soviet research continued, but publications were mainly in Russian or Georgian languages and were unavailable internationally for many years.

Their use has continued since the end of the Cold War in Georgia and elsewhere in Central and Eastern Europe. The first regulated, randomized, double-blind clinical trial was reported in the Journal of Wound Care in June 2009, which evaluated the safety and efficacy of a bacteriophage cocktail to treat infected venous leg ulcers in human patients. The FDA approved the study as a Phase I clinical trial. The study's results demonstrated the safety of therapeutic application of bacteriophages but did not show efficacy. The authors explain that the use of certain chemicals that are part of standard wound care (e.g. lactoferrin or silver) may have interfered with bacteriophage viability. Another controlled clinical trial in Western Europe (treatment of ear infections caused by *Pseudomonas aeruginosa*) was reported shortly after in the journal Clinical Otolaryngology in August 2009. The study concludes that bacteriophage preparations were safe and effective for treatment of chronic ear infections in humans. Additionally, there have been numerous animal and other experimental clinical trials evaluating the efficacy of bacteriophages for various diseases, such as infected burns and wounds, and cystic fibrosis associated lung infections, among others. Meanwhile, bacteriophage researchers are developing engineered viruses to overcome antibiotic resistance, and engineering the phage genes responsible for coding enzymes which degrade the biofilm matrix, phage structural proteins and also enzymes responsible for lysis of bacterial cell wall.

D'Herelle "quickly learned that bacteriophages are found wherever bacteria thrive: in sewers, in rivers that catch waste runoff from pipes, and in the stools of convalescent patients." This includes rivers traditionally thought to have healing powers, including India's Ganges River.

Replication

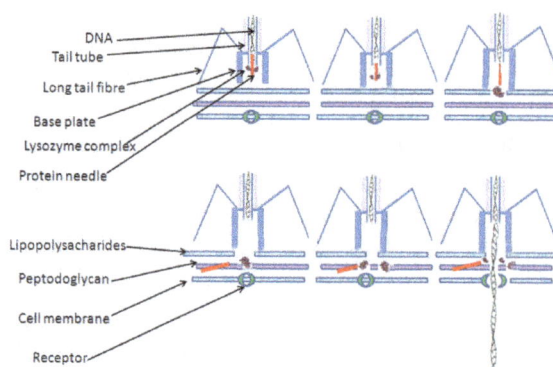

Diagram of the DNA injection process

Bacteriophages may have a lytic cycle or a lysogenic cycle, and a few viruses are capable of carrying out both. With *lytic phages* such as the T4 phage, bacterial cells are broken open (lysed) and destroyed after immediate replication of the virion. As soon as the cell is destroyed, the phage progeny can find new hosts to infect. Lytic phages are more suitable for phage therapy. Some lytic phages undergo a phenomenon known as lysis inhibition, where completed phage progeny will not immediately lyse out of the cell if extracellular phage concentrations are high. This mechanism is not identical to that of temperate phage going dormant and is usually temporary.

In contrast, the *lysogenic cycle* does not result in immediate lysing of the host cell. Those phages able to undergo lysogeny are known as temperate phages. Their viral genome will integrate with host DNA and replicate along with it relatively harmlessly, or may even become established as a plasmid. The virus remains dormant until host conditions deteriorate, perhaps due to depletion of nutrients; then, the endogenous phages (known as prophages) become active. At this point they initiate the reproductive cycle, resulting in lysis of the host cell. As the lysogenic cycle allows the host cell to continue to survive and reproduce, the virus is replicated in all of the cell's offspring. An example of a bacteriophage known to follow the lysogenic cycle and the lytic cycle is the phage lambda of *E. coli*.

Sometimes prophages may provide benefits to the host bacterium while they are dormant by adding new functions to the bacterial genome in a phenomenon called lysogenic conversion. Examples are the conversion of harmless strains of *Corynebacterium diphtheriae* or *Vibrio cholerae* by bacteriophages to highly virulent ones, which cause diphtheria or cholera, respectively. Strategies to combat certain bacterial infections by targeting these toxin-encoding prophages have been proposed.

Attachment and Penetration

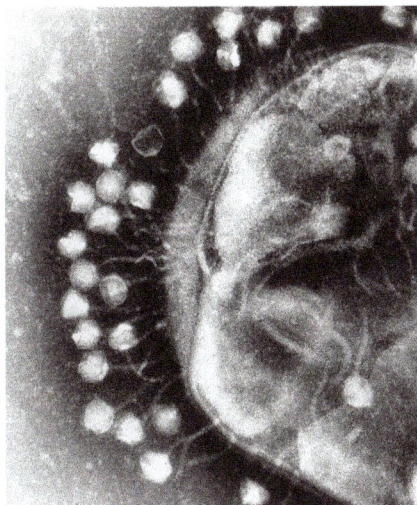

In this electron micrograph of bacteriophages attached to a bacterial cell, the viruses are the size and shape of coliphage T1.

To enter a host cell, bacteriophages attach to specific receptors on the surface of bacteria, including lipopolysaccharides, teichoic acids, proteins, or even flagella. This specificity means a bacteriophage can infect only certain bacteria bearing receptors to which they can bind, which in turn determines the phage's host range. Host growth conditions also influence the ability of the phage to attach and invade them. As phage virions do not move independently, they must rely on random encounters with the right receptors when in solution (blood, lymphatic circulation, irrigation, soil water, etc.).

Myovirus bacteriophages use a hypodermic syringe-like motion to inject their genetic material into the cell. After making contact with the appropriate receptor, the tail fibers flex to bring the base plate closer to the surface of the cell; this is known as reversible binding. Once attached completely, irreversible binding is initiated and the tail contracts, possibly with the help of ATP present in the tail, injecting genetic material through the bacterial membrane. Podoviruses lack an elongated tail sheath similar to that of a myovirus, so they instead use their small, tooth-like tail fibers enzymatically to degrade a portion of the cell membrane before inserting their genetic material.

Synthesis of Proteins and Nucleic Acid

Within minutes, bacterial ribosomes start translating viral mRNA into protein. For RNA-based phages, RNA replicase is synthesized early in the process. Proteins modify the bacterial RNA polymerase so it preferentially transcribes viral mRNA. The host's normal synthesis of proteins and nucleic acids is disrupted, and it is forced to manufacture viral products instead. These products go on to become part of new virions within the cell, helper proteins that help assemble the new virions, or proteins involved in cell lysis. Walter Fiers (University of Ghent, Belgium) was the first to establish the complete nucleotide sequence of a gene (1972) and of the viral genome of bacteriophage MS2 (1976).

Virion Assembly

In the case of the T4 phage, the construction of new virus particles involves the assistance of helper proteins. The base plates are assembled first, with the tails being built upon them afterward. The head capsids, constructed separately, will spontaneously assemble with the tails. The DNA is packed efficiently within the heads. The whole process takes about 15 minutes.

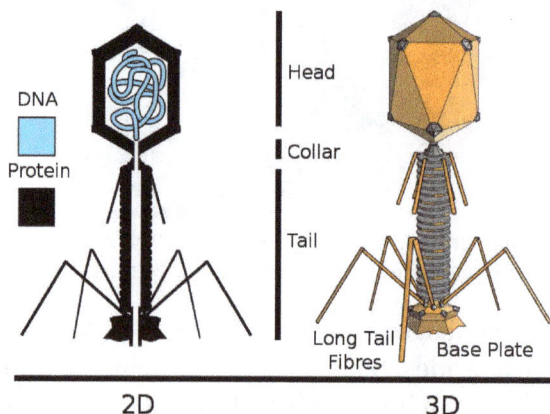

Diagram of a typical tailed bacteriophage structure

Release of Virions

Phages may be released via cell lysis, by extrusion, or, in a few cases, by budding. Lysis, by tailed phages, is achieved by an enzyme called endolysin, which attacks and breaks down the cell wall peptidoglycan. An altogether different phage type, the filamentous phages, make the host cell continually secrete new virus particles. Released virions are described as free, and, unless defective, are capable of infecting a new bacterium. Budding is associated with certain *Mycoplasma* phages. In contrast to virion release, phages displaying a lysogenic cycle do not kill the host but, rather, become long-term residents as prophage.

Genome Structure

Given the millions of different phages in the environment, phages' genomes come in a variety of forms and sizes. RNA phage such as MS2 have the smallest genomes of only a few kilobases. However, some DNA phages such as T4 may have large genomes with hundreds of genes; the size and shape of the capsid varies along with the size of the genome.

Bacteriophage genomes can be highly mosaic, i.e. the genome of many phage species appear to be composed of numerous individual modules. These modules may be found in other phage species in different arrangements. Mycobacteriophages – bacteriophages with mycobacterial hosts – have provided excellent examples of this mosaicism. In these mycobacteriophages, genetic assortment may be the result of repeated instances of site-specific recombination and illegitimate recombination (the result of phage genome acquisition of bacterial host genetic sequences). It should be noted, however, that evolutionary mechanisms shaping the genomes of bacterial viruses vary between different families and depend on the type of the nucleic acid, characteristics of the virion structure, as well as the mode of the viral life cycle.

Systems Biology

Phages often have dramatic effects on their hosts. As a consequence, the transcription pattern of the infected bacterium may change considerably. For instance, infection of *Pseudomonas aeruginosa* by the temperate phage PaP3 changed the expression of 38% (2160/5633) of its host's genes. Many of these effects are probably indirect, hence the challenge becomes to identify the direct interactions among bacteria and phage.

Several attempts have been made to map Protein–protein interactions among phage and their host. For instance, bacteriophage lambda was found to interact with its host E. coli by 31 interactions. However, a large-scale study revealed 62 interactions, most of which were new. Again, the significance of many of these interactions remains unclear, but these studies suggest that there are most likely several key interactions and many indirect interactions whose role remains uncharacterized.

In the Environment

Metagenomics has allowed the in-water detection of bacteriophages that was not possible previously.

Bacteriophages have also been used in hydrological tracing and modelling in river systems, especially where surface water and groundwater interactions occur. The use of phages is preferred to the more conventional dye marker because they are significantly less absorbed when passing through ground waters and they are readily detected at very low concentrations. Non-polluted water may contain ca. 2×10^8 bacteriophages per mL.

Bacteriophages are thought to extensively contribute to horizontal gene transfer in natural environments, principally via transduction but also via transformation. Metagenomics-based studies have also revealed that viromes from a variety of environments harbor antibiotic resistance genes, including those that could confer multidrug resistance.

Other Areas of use

Since 2006, the United States Food and Drug Administration (FDA) and United States Department of Agriculture (USDA) have approved several bacteriophage products. LMP-102 (Intralytix) was approved for treating ready-to-eat (RTE) poultry and meat products. In that same year, the FDA approved LISTEX (developed and produced by Micreos) using bacteriophages on cheese to kill *Listeria monocytogenes* bacteria, giving them generally recognized as safe (GRAS) status. In July 2007, the same bacteriophage were approved for use on all food products. In 2011 USDA confirmed that LISTEX is a clean label processing aid and is included in USDA. Research in the field of food safety is continuing to see if lytic phages are a viable option to control other foodborne pathogens in various food products.

In 2011 the FDA cleared the first bacteriophage-based product for in vitro diagnostic use. The KeyPath MRSA/MSSA Blood Culture Test uses a cocktail of bacteriophage to detect *Staphylococcus aureus* in positive blood cultures and determine methicillin resistance or susceptibility. The test returns results in about 5 hours, compared to 2–3 days for standard microbial identification and susceptibility test methods. It was the first accelerated antibiotic susceptibility test approved by the FDA.

Government agencies in the West have for several years been looking to Georgia and the former Soviet Union for help with exploiting phages for counteracting bioweapons and toxins, such as anthrax and botulism. Developments are continuing among research groups in the US. Other uses include spray application in horticulture for protecting plants and vegetable produce from decay and the spread of bacterial disease. Other applications for bacteriophages are as biocides for environmental surfaces, e.g., in hospitals, and as preventative treatments for catheters and medical devices before use in clinical settings. The technology for phages to be applied to dry surfaces, e.g., uniforms, curtains, or even sutures for surgery now exists. Clinical trials reported in *Clinical Otolaryngology* show success in veterinary treatment of pet dogs with otitis.

Phage display is a different use of phages involving a library of phages with a variable peptide linked to a surface protein. Each phage's genome encodes the variant of the protein displayed on its surface (hence the name), providing a link between the peptide variant and its encoding gene. Variant phages from the library can be selected through their binding affinity to an immobilized molecule (e.g., botulism toxin) to neutralize it. The bound, selected phages can be multiplied by reinfecting a susceptible bacterial strain, thus allowing them to retrieve the peptides encoded in them for further study.

The SEPTIC bacterium sensing and identification method uses the ion emission and its dynamics during phage infection and offers high specificity and speed for detection.

Phage-ligand technology makes use of proteins, which are identified from bacteriophages, characterized and recombinantly expressed for various applications such as binding of bacteria and bacterial components (e.g. endotoxin) and lysis of bacteria.

Bacteriophages are also important model organisms for studying principles of evolution and ecology.

Model Bacteriophages

The following bacteriophages are extensively studied:

- λ phage

- T2 phage

- T4 phage (169 kbp genome, 200 nm long)

- T7 phage

- T12 phage

- R17 phage

- M13 phage

- MS2 phage (23–28 nm in size)

- G4 phage

- P1 phage

- Enterobacteria phage P2

- P4 phage

- Phi X 174 phage

- N4 phage

- Pseudomonas phage Φ6

- Φ29 phage

- 186 phage

References

- Zhang XHD (2008). "Novel analytic criteria and effective plate designs for quality control in genome-scale RNAi screens". Journal of Biomolecular Screening. 13 (5): 363–77. PMID 18567841. doi:10.1177/1087057108317062

- Zhang XHD (2011). Optimal High-Throughput Screening: Practical Experimental Design and Data Analysis for Genome-scale RNAi Research. Cambridge University Press. ISBN 978-0-521-73444-8

- Quantum Materials Corporation and the Access2Flow Consortium (2011). "Quantum materials corp achieves milestone in High Volume Production of Quantum Dots". Retrieved 7 July 2011

- Brideau C, Gunter G, Pikounis B, Liaw A (2003). "Improved statistical methods for hit selection in high-throughput screening". Journal of Biomolecular Screening. 8 (6): 634–47. PMID 14711389. doi:10.1177/1087057103258285

- Chaga GS (2008). "Antibody arrays for determination of relative protein abundances". Methods Mol. Biol. Methods in Molecular Biology. 441: 129–51. ISBN 978-1-58829-679-5. PMID 18370316. doi:10.1007/978-1-60327-047-2_9

- MFTTech (24 March 2015). "LG Electronics Partners with Dow to Commercialize LGs New Ultra HD TV with Quantum Dot Technology". Retrieved 9 May 2015

- Melton, Lisa (2004). "Protein arrays: Proteomics in multiplex". Nature. 429 (6987): 101–107. ISSN 0028-0836. doi:10.1038/429101a

- Jiang, W., Mao, Y. Q., Huang, R., Duan, C., Xi, Y., Yang, K., & Huang, R. P. (2014). "Protein expression profiling by antibody array analysis with use of dried blood spot samples on filter paper". Journal of immunological methods. 403 (1): 79-86. PMID 24287424

- Scott JS, Barbas CF III, Burton, DA (2001). Phage Display: A Laboratory Manual. Plainview, N.Y: Cold Spring Harbor Laboratory Press. ISBN 0-87969-740-7

- Mangolini, L.; Kortshagen, U. (2007). "Plasma-assisted synthesis of silicon nanocrystal inks". Advanced Materials. 19 (18): 2513–2519. doi:10.1002/adma.200700595

- Leatherdale, C. A.; Woo, W. -K.; Mikulec, F. V.; Bawendi, M. G. (2002). "On the Absorption Cross Section of CdSe Nanocrystal Quantum Dots". The Journal of Physical Chemistry B. 106 (31): 7619–7622. doi:10.1021/jp025698c

- Mc Grath S and van Sinderen D (editors). (2007). Bacteriophage: Genetics and Molecular Biology (1st ed.). Caister Academic Press. ISBN 978-1-904455-14-1.

- LaFave, T. Jr. (2013). "Correspondences between the classical electrostatic Thomson Problem and atomic electronic structure". Journal of Electrostatics. 71 (6): 1029–1035. doi:10.1016/j.elstat.2013.10.001

- Bednarek, S.; Szafran, B. & Adamowski, J. (1999). "Many-electron artificial atoms". Phys. Rev. B. 59 (20): 13036–13042. Bibcode:1999PhRvB..5913036B. doi:10.1103/PhysRevB.59.13036

4

Label-free Detection: An Overview

Label-free detection is a detection technique used in protein microarrays. Microcantilever, scanning kelvin nanoprobe, atomic force microscopy, etc. are some label-free detection techniques. This chapter is a compilation of the various branches of label-free detection that form an integral part of the broader subject matter.

Label-free Detection Approaches

Over the last two decades detection techniques in proteomics have experienced phenomenal achievements with introduction of different ultra-sensitive detection techniques, which can selectively detect target analytes from even complex biological samples. Additionally, quite a few detection techniques are proficient for multiplexed detection, which is very useful for high-throughput proteomics applications, particularly in protein/antibody microarrays. There are two major detection approaches used in protein microarrays; label-based and label-free. In label-free detection approaches inherent properties of the query molecules, like mass and dielectric property are measured. This eliminates any requirement or interference due to the presence of tagging molecules. Recently, label-free detection methods are gaining recognition due to their easy operation procedure, real-time detection, exclusion of the need of secondary reactants and prolonged labeling practice.

Fluorescence Spectroscopy

Fluorescence spectroscopy (also known as fluorometry or spectrofluorometry) is a type of electromagnetic spectroscopy that analyzes fluorescence from a sample. It involves using a beam of light, usually ultraviolet light, that excites the electrons in molecules of certain compounds and causes them to emit light; typically, but not necessarily, visible light. A complementary technique is absorption spectroscopy. In the special case of single molecule fluorescence spectroscopy, intensity fluctuations from the emitted light are measured from either single fluorophores, or pairs of fluorophores.

Devices that measure fluorescence are called fluorometers.

Theory

Molecules have various states referred to as energy levels. Fluorescence spectroscopy is primarily concerned with electronic and vibrational states. Generally, the species being examined has a ground electronic state (a low energy state) of interest, and an excited electronic state of higher energy. Within each of these electronic states there are various vibrational states.

In fluorescence, the species is first excited, by absorbing a photon, from its ground electronic state to one of the various vibrational states in the excited electronic state. Collisions with other molecules cause the excited molecule to lose vibrational energy until it reaches the lowest vibrational state of the excited electronic state. This process is often visualized with a Jablonski diagram.

The molecule then drops down to one of the various vibrational levels of the ground electronic state again, emitting a photon in the process. As molecules may drop down into any of several vibrational levels in the ground state, the emitted photons will have different energies, and thus frequencies. Therefore, by analysing the different frequencies of light emitted in fluorescent spectroscopy, along with their relative intensities, the structure of the different vibrational levels can be determined.

For atomic species, the process is similar; however, since atomic species do not have vibrational energy levels, the emitted photons are often at the same wavelength as the incident radiation. This process of re-emitting the absorbed photon is "resonance fluorescence" and while it is characteristic of atomic fluorescence, is seen in molecular fluorescence as well.

In a typical fluorescence (emission) measurement, the excitation wavelength is fixed and the detection wavelength varies, while in a fluorescence excitation measurement the detection wavelength is fixed and the excitation wavelength is varied across a region of interest. An emission map is measured by recording the emission spectra resulting from a range of excitation wavelengths and combining them all together. This is a three dimensional surface data set: emission intensity as a function of excitation and emission wavelengths, and is typically depicted as a contour map.

Instrumentation

Two general types of instruments exist: filter fluorometers that use filters to isolate the incident light and fluorescent light and spectrofluorometers that use a diffraction gratingmonochromators to isolate the incident light and fluorescent light.

Both types use the following scheme: the light from an excitation source passes through a filter or monochromator, and strikes the sample. A proportion of the incident light is absorbed by the sample, and some of the molecules in the sample fluoresce. The fluorescent light is emitted in all directions. Some of this fluorescent light passes through a second filter or monochromator and reaches a detector, which is usually placed at 90° to the incident light beam to minimize the risk of transmitted or reflected incident light reaching the detector.

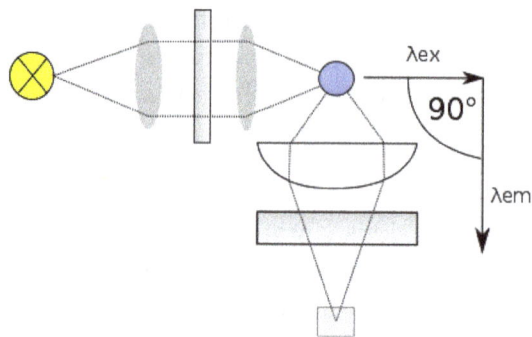

A simplistic design of the components of a fluorimeter

Various light sources may be used as excitation sources, including lasers, LED, and lamps; xenon arcs and mercury-vapor lamps in particular. A laser only emits light of high irradiance at a very narrow wavelength interval, typically under 0.01 nm, which makes an excitation monochromator or filter unnecessary. The disadvantage of this method is that the wavelength of a laser cannot be changed by much. A mercury vapor lamp is a line lamp, meaning it emits light near peak wavelengths. By contrast, a xenon arc has a continuous emission spectrum with nearly constant intensity in the range from 300-800 nm and a sufficient irradiance for measurements down to just above 200 nm.

Filters and/or monochromators may be used in fluorimeters. A monochromator transmits light of an adjustable wavelength with an adjustable tolerance. The most common type of monochromator utilizes a diffraction grating, that is, collimated light illuminates a grating and exits with a different angle depending on the wavelength. The monochromator can then be adjusted to select which wavelengths to transmit. For allowing anisotropy measurements, the addition of two polarization filters is necessary: One after the excitation monochromator or filter, and one before the emission monochromator or filter.

As mentioned before, the fluorescence is most often measured at a 90° angle relative to the excitation light. This geometry is used instead of placing the sensor at the line of the excitation light at a 180° angle in order to avoid interference of the transmitted excitation light. No monochromator is perfect and it will transmit some stray light, that is, light with other wavelengths than the targeted. An ideal monochromator would only transmit light in the specified range and have a high wavelength-independent transmission. When measuring at a 90° angle, only the light scattered by the sample causes stray light. This results in a better signal-to-noise ratio, and lowers the detection limit by approximately a factor 10000, when compared to the 180° geometry. Furthermore, the fluorescence can also be measured from the front, which is often done for turbid or opaque samples.

The detector can either be single-channeled or multichanneled. The single-channeled detector can only detect the intensity of one wavelength at a time, while the multichanneled detects the intensity of all wavelengths simultaneously, making the emission monochromator or filter unnecessary. The different types of detectors have both advantages and disadvantages.

The most versatile fluorimeters with dual monochromators and a continuous excitation light source can record both an excitation spectrum and a fluorescence spectrum. When measuring fluorescence spectra, the wavelength of the excitation light is kept constant, preferably at a wavelength of high absorption, and the emission monochromator scans the spectrum. For measuring excitation spectra, the wavelength passing though the emission filter or monochromator is kept constant and the excitation monochromator is scanning. The excitation spectrum generally is identical to the absorption spectrum as the fluorescence intensity is proportional to the absorption.

Analysis of Data

At low concentrations the fluorescenceintensity will generally be proportional to the concentration of the fluorophore.

Unlike in UV/visible spectroscopy, 'standard', device independent spectra are not easily attained. Several factors influence and distort the spectra, and corrections are necessary to attain 'true', i.e. machine-independent, spectra. The different types of distortions will here be classified as being either instrument- or sample-related. Firstly, the distortion arising from the instrument is discussed. As a start, the light source intensity and wavelength characteristics varies over time during each experiment and between each experiment. Furthermore, no lamp has a constant intensity at all wavelengths. To correct this, a beam splitter can be applied after the excitation monochromator or filter to direct a portion of the light to a reference detector.

Additionally, the transmission efficiency of monochromators and filters must be taken into account. These may also change over time. The transmission efficiency of the monochromator also varies depending on wavelength. This is the reason that an optional reference detector should be placed after the excitation monochromator or filter. The percentage of the fluorescence picked up by the detector is also dependent upon the system. Furthermore, the detector quantum efficiency, that is, the percentage of photons detected, varies between different detectors, with wavelength and with time, as the detector inevitably deteriorates.

Two other topics that must be considered include the optics used to direct the radiation and the means of holding or containing the sample material (called a cuvette or cell). For most UV, visible, and NIR measurements the use of precision quartz cuvettes is necessary. In both cases, it is important to select materials that have relatively little absorption in the wavelength range of interest. Quartz is ideal because it transmits from 200 nm-2500 nm; higher grade quartz can even transmit up to 3500 nm, whereas the absorption properties of other materials can mask the fluorescence from the sample.

Correction of all these instrumental factors for getting a 'standard' spectrum is a tedious process, which is only applied in practice when it is strictly necessary. This is the case when measuring the quantum yield or when finding the wavelength with the highest emission intensity for instance.

As mentioned earlier, distortions arise from the sample as well. Therefore, some aspects of the sample must be taken into account too. Firstly, photodecomposition may decrease the intensity of fluorescence over time. Scattering of light must also be taken into account. The most significant types of scattering in this context are Rayleigh and Raman scattering. Light scattered by Rayleigh scattering has the same wavelength as the incident light, whereas in Raman scattering the scattered light changes wavelength usually to longer wavelengths. Raman scattering is the result of a virtual electronic state induced by the excitation light. From this virtual state, the molecules may relax back to a vibrational level other than the vibrational ground state. In fluorescence spectra, it is always seen at a constant wavenumber difference relative to the excitation wavenumber e.g. the peak appears at a wavenumber 3600 cm^{-1} lower than the excitation light in water.

Other aspects to consider are the inner filter effects. These include reabsorption. Reabsorption happens because another molecule or part of a macromolecule absorbs at the wavelengths at which the fluorophore emits radiation. If this is the case, some or all of the photons emitted by the fluorophore may be absorbed again. Another inner filter effect occurs because of high concentrations of absorbing molecules, including the fluorophore. The result is that the intensity of the excitation light is not constant throughout the solution. Resultingly, only a small percentage of the excitation light reaches the fluorophores that are visible for the detection system. The inner filter effects change the spectrum and intensity of the emitted light and they must therefore be considered when analysing the emission spectrum of fluorescent light.

Tryptophan Fluorescence

The fluorescence of a folded protein is a mixture of the fluorescence from individual aromatic residues. Most of the intrinsic fluorescence emissions of a folded protein are due to excitation of tryptophan residues, with some emissions due to tyrosine and phenylalanine; but disulfide bonds also have appreciable absorption in this wavelength range. Typically, tryptophan has a wavelength of maximum absorption of 280 nm and an emission peak that is solvatochromic, ranging from ca. 300 to 350 nm depending in the polarity of the local environment Hence, protein fluorescence may be used as a diagnostic of the conformational state of a protein. Furthermore, tryptophan fluorescence is strongly influenced by the proximity of other residues (*i.e.*, nearby *protonated* groups such as Asp or Glu can cause quenching of Trp fluorescence). Also, energy transfer between tryptophan and the other fluorescent amino acids is possible, which would affect the analysis, especially in cases where the Förster acidic approach is taken. In addition, tryptophan is a relatively rare amino acid; many proteins contain only one or a few tryptophan residues. Therefore, tryptophan fluorescence can be a very sensitive measurement of the conformational state of individual tryptophan residues. The advantage compared to extrinsic probes is that the protein itself is not changed. The use of intrinsic fluorescence for the study of protein conformation is in practice limited to cases with few (or perhaps only one) tryptophan residues, since each experiences a different local environment, which gives rise to different emission spectra.

Tryptophan is an important intrinsic fluorescent probe (amino acid), which can be used to estimate the nature of microenvironment of the tryptophan. When performing experiments with denaturants, surfactants or other amphiphilic molecules, the micro-environment of the tryptophan might change. For example, if a protein containing a single tryptophan in its 'hydrophobic' core is denatured with increasing temperature, a red-shifted emission spectrum will appear. This is due to the exposure of the tryptophan to an aqueous environment as opposed to a hydrophobic protein interior. In contrast, the addition of a surfactant to a protein which contains a tryptophan which is exposed to the aqueous solvent will cause a blue-shifted emission spectrum if the tryptophan is embedded in the surfactant vesicle or micelle. Proteins that lack tryptophan may be coupled to a fluorophore.

With fluorescence excitation at 295 nm, the tryptophan emission spectrum is dominant over the weaker tyrosine and phenylalanine fluorescence.

Applications

Fluorescence spectroscopy is used in, among others, biochemical, medical, and chemical research fields for analyzing organic compounds. There has also been a report of its use in differentiating malignant skin tumors from benign.

Atomic Fluorescence Spectroscopy (AFS) techniques are useful in other kinds of analysis/measurement of a compound present in air or water, or other media, such as CVAFS which is used for heavy metals detection, such as mercury.

Additionally, Fluorescence spectroscopy can be adapted to the microscopic level using microfluorimetry.

In analytical chemistry, fluorescence detectors are used with HPLC.

Label-based Detection Techniques

With rapid advancements in gel-free proteomics techniques, particularly protein microarrays, the need for improved detection systems has been imperative. Label-based detection systems have taken rapid strides to satisfy this demand with significant improvements in sensitivity, multiplexing capability and reproducibility.

Different Label-Free Detection Techniques Commonly used in Proteomics.

Development of reliable, sensitive and high-throughput label-free detection techniques has become imperative for proteomic studies due to drawbacks associated with label-based technologies Label-free detection methods, which monitor inherent properties of the query molecule, promise to simplify bioassays.

Overview of Label-free Techniques

Label-free detection: Label-free detection techniques monitor inherent properties of the query molecules such as mass, optical and dielectric properties. Unlike label-based detection methods, these techniques avoid any tagging of the query molecules thereby preventing changes in structure and function. They do not involve laborious procedures but have their own pitfalls such as sensitivity and specificity issues.

Surface Plasmon Resonance-based Techniques

i) Surface plasmon resonance (SPR): Detects any change in refractive index of material at the interface between metal surface and the ambient medium.

ii) Surface plasmon resonance imaging (SPRi): Image reflected by polarized light at fixed angle detected.

iii) Nanohole array: Light transmission of specific wavelength enhanced by coupling of surface plasmons on both sides of metal surface with periodic nanoholes.

Ellipsometry-based techniques

i) Ellipsometry: Change in polarization state of reflected light arising due to changes in dielectric property or refractive index of surface material measured.

ii) Oblique incidence reflectivity difference (OI-RD): Variation of ellipsometry that monitors harmonics of modulated photocurrents under nulling conditions.

Interference-based techniques: Interferometry is based on the principle of transformation of phase differences of wave fronts into readily recordable intensity fluctuations known as interference fringes. The various detection strategies that make use of this principle include:

Spectral reflectance imaging biosensor (SRIB): Changes in optical index due to capture of molecules on the array surface detected using optical wave interference.

ii) Biological compact disc (BioCD): Local interferometry i.e. transformation of phase differences of wave fronts into observable interference fringes, used for detection of protein capture.

iii) Arrayed imaging reflectometry (AIR): Destructive interference of polarized light reflected from silicon substrate captured and used for detection.

Electrochemical impedance spectroscopy (EIS) -aptamer array: Aptamers are short single-stranded oligonucleotides that are capable of binding to a wide range of target biomolecules. EIS combined with aptamer arrays can offer a highly sensitive label-free detection technique.

Atomic force microscopy (AFM): Vertical or horizontal deflections of cantilever measured by high-resolution scanning probe microscope, thereby providing significant information about surface features.

Enthalpy array: Thermodynamics and kinetics of molecular interactions measured in small sample volumes without any need for immobilization or labelling of reactants.

Scanning Kelvin nanoprobe (SKN): A non-contact technique that does not require specialized vacuum or fluid cell; SKN detects regional variations in surface potential across the substrate of interest caused due to molecular interactions.

Microcantilever: These are thin, silicon-based, gold-coated surfaces that hang from a solid support. Bending of cantilever due to surface adsorption is detected either electrically by metal oxide semiconductor field effect transistors or optically by changes in angle of reflection.

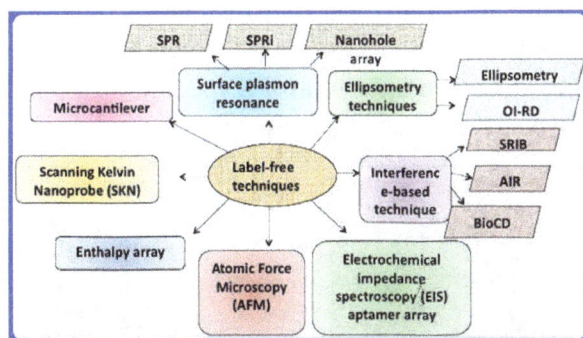

An overview of label-free techniques

(1A.) SPR and related techniques

Surface plasmon resonance (SPR) is a label-free technique that measures variations in refractive index of the dielectric layer adjoining to the sensor surface as a result of the adsorption or desorption of molecules. SPR provides real-time measurements of alteration in refractive index in the locality of a surface (Shankaran et al., 2007). The variation in reflection intensity with respect to incident angle before and after binding of the target molecule is shown as sensorgram.

In SPR imaging (SPRi) a spatially resolved measuring device is introduced in SPR set-up (Ladd et al., 2009). SPR and SPRi are suitable for instantaneous label-free analysis of several biomolecular interactions in a quick and HT style. SPR-based biosensors capable of detecting very minute amounts of target analytes with high selectivity are promising for discovery of disease biomarkers.

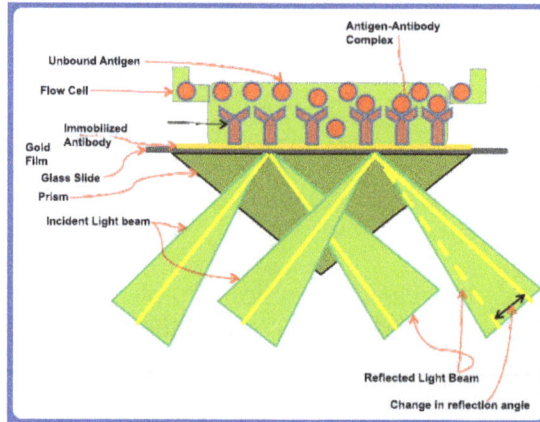

Variations in the refractive index of the medium directly in contact with sensor surface is measured in SPR.

Apart from SPR and SPRi, nanohole arrays are considered as a advantageous label-free approach for biosensing, since they require plain optical alignment and simple miniaturization, and offer high accuracy, robustness, increased fluorescent signal, multiplexing and collinear optical detection. If prearranged arrays of nanoscale holes are designed in a metal film, unusual optical transmission characteristics at resonant wavelengths are observed. Surface plasmons (SPs) are excited on both sides of metal surface. It increases the light transmission for a specific wavelength and makes nanohole arrays a prospective surface-based biosensor.

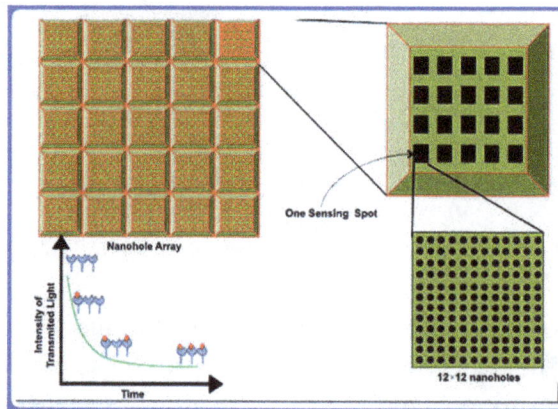

Nanohole arrays are label-free approach for biosensing, which require plain optical alignment and simple miniaturization, and offer high accuracy, robustness, increased fluorescent signal, multiplexing and collinear optical detection. If prearranged arrays of nanoscale holes are designed in a metal film, unusual optical transmission characteristics at resonant wavelengths are observed.

(b) Ellipsometry-based techniques

Ellipsometry-based label-free detection methods measure the polarization state of the reflected light, which is changed when dielectric property or refractive index of the sample surface is altered. In imaging ellipsometry microscopy the CCD camera are coupled with ellipsometer (Jin et al., 2004). If microfluidic system is coupled with imaging ellipsometry multiple advantages such as high automation, less sample consumption, fast detection and HT assays with superior sensitivity can be achieved. Ellipsometry-based techniques are useful for studying kinetics of biomolecular interactions, hormonal activity, detection of microorganisms, and quantification of competitive adsorption of protein

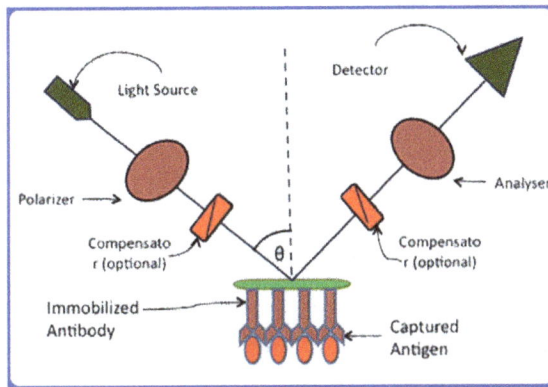

Basic principle behind ellipsometry. A photodetector is used to monitor the intensity of the reflected light.

Another form of ellipsometry; oblique incidence reflectivity difference (OI-RD), in which the harmonics of modulated photocurrents are measured under appropriate nulling conditions, is used as a label-free detection platform in proteomics. Variation in thickness and/or dielectric response as a result of biomolecular interactions generates a detectable OI-RD signal.

(c) Interference-based techniques

Interference-based techniques detect optical phase difference as a result of biomolecular mass addition. There are quite a few potential interferometric techniques such as spectral reflectance imaging biosensor (SRIB), dual-channel biosensor, SPR interferometry, on chip interferometric backscatter detection, porous silicon-based optical interferometric biosensor, biological compact disc (BioCD) and spinning disc interferometry, which are very promising for label-free detection of biomolecules. Biochemical and functional analysis of proteins are also possible using interference-based label-free detection methods.

Spectral reflectance imaging biosensor (SRIB) is the most promising interference-based label-free detection method, which monitors alterations in the optical index due to the capture of biological material on the sensor surface (Ozkumur et al., 2008). SRIB directly monitors primary molecular binding interactions with high sensitivity. Back-scattering interferometry (BSI) is another promising platform for studying label-free molecular interactions within very small amount of samples (Bornhop et al., 2007). It can quantify a wide dynamic range (Kd spanning six decades) of molecular interactions in free solution and very compatible for multiplexing.

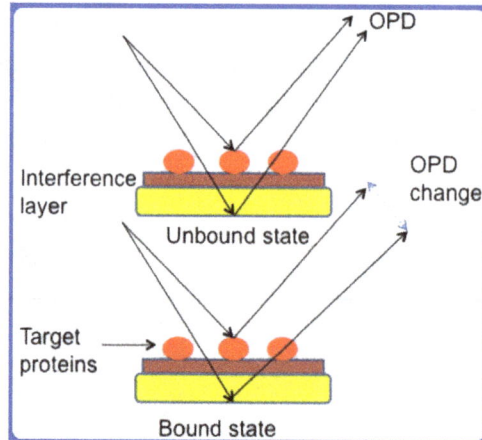

Interferometric techniques measure the phase differences of the wave fronts and convert it into observable visible intensity fluctuation known as Interference fringes.

(d) Scanning Kelvin nanoprobe

The Kelvin probe force microscope (KPFM) measures local changes in surface potential across a substrate (generally gold). KPFM is very promising for label-free biomolecular label-free detection and offers a number of advantages. It is a non-contact technique; therefore specialized vacuum or fluid cell is not required. High-speed screening is possible with KPFM while maintaining the signal fidelity. Another significant aspect of KPFM technology is its ability to analyse high-density arrays (Sinensky et al., 2007). This label-free detection technique also has ability to reduce noise by decreasing the non-specific binding of biomolecules.

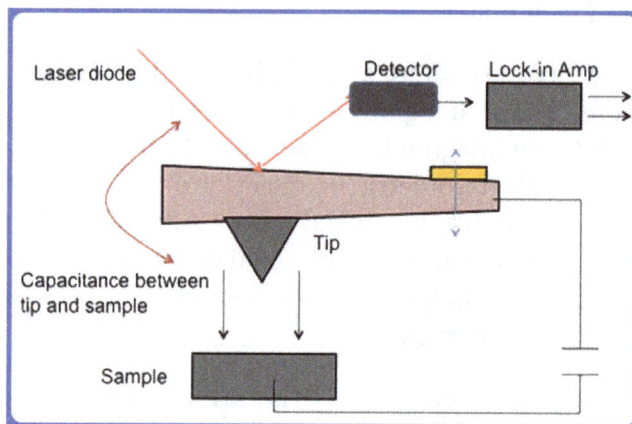

Working principle of Scanning Kelvin nanoprobe. Variations in work function and surface potential occurred due to molecular interactions are measured using SKN.

(e) Microcantilever

Microcantilevers are silicon-based, gold-coated, thin (1 mm) surfaces, horizontally attached to a solid support (Braun et al., 2009). Binding of biomolecules on the cantilevers bends them and level of bending is measured optically or electrically for label-free detection. Analysis of thermodynamics of protein–protein and other biomolecular interactions, detection of cancer markers, antigen–antibody interactions can be performed using microcantilever-based label-free sensors.

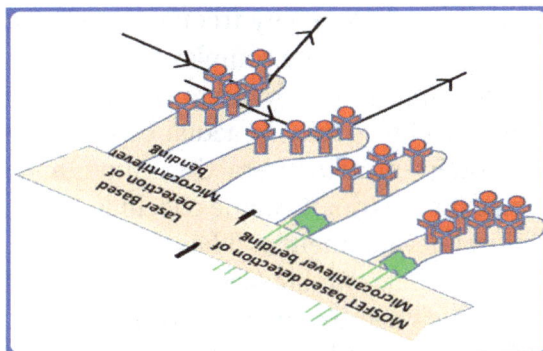

Working principle of microcantilever. The interaction of query molecules with immobilized target molecules leads to bending of microcantilevers and changes the resonant frequency.

II. Application of Label- Free Techniques in Proteomics

Quite a few label-free approaches, like SPR, SPRi, interference-based techniques, microcantilever etc. are considered as impending platforms for studying biomolecular interactions and detection of disease biomarkers.

III. Challenges

Regardless of rapid advancement in field of label-free proteomics achieved with introduction of new and versatile technologies, label-free detection approaches have several limitations as well. Both label-free and label-based detection methods have their own advantages and limitations (Ray et al., 2010; Chandra et al., 2011). Although, label-free detections techniques are very promising and potential candidates for real-time measurements of low-abundance analytes and protein-protein interactions, issues regarding sensitivity and specificity remain to be explored further. Additionally, costly fabrication techniques, morphological anomalies of sample spots and insufficient knowledge regarding the exact working principles of the label-free biosensors often restrict their use for practical clinical applications. Label-free measurements have capabilities of detection of low-abundance analytes and protein-protein interactions; but further improvement of specificity and sensitivity is required when complex body fluids have to be analyzed rather than simple buffer solutions commonly used in most of the proof-of-principle experiments. For making label-free sensors popular in routine clinical applications, cost-effective fabrication techniques are required to be developed, and mechanism of working principle of label-free detection approaches need to be explored further.

Label-free detection techniques are definitely attractive for the large-scale, real-time analysis of protein–protein and other biomolecular interactions and measurement of concentrations of multiple target molecules in HT manner. Such extremely sensitive, fast, label-free detection approaches are useful in various applied fields including pharmaceutical analysis, screening of potential drug molecules, cellular detection, characterization of biomolecules, detection of disease markers and environmental monitoring. Coupling of microarrays and label-free techniques is emerging rapidly and found to be highly effective in detection of extremely low-abundance analytes in buffer solutions. Nonetheless, sensitivity and specificity frequently become the prime limitation for label-free detection methods when very complex biological samples are concerned. Hitherto, label-free detection approaches found to be efficient in analysis of antigen–antibody interactions, but it will be useful in actual bedside applications in clinics, if they can detect multiple protein–protein interactions simultaneously with similar efficacy. Considering the present scenario, it can be concluded that label-free proteomics is still at a premature stage of development and has shown promises mainly for targeted detection of known protein parkers; however, it has not contributed effectively in discovery of new markers which can be directly translated in clinics. It is anticipated that with efforts from different research groups world-wide, the field of label-free proteomics will turn into more robust, sensitive, fast, cost-effective and overcome existing limitations.

Surface Plasmon Resonance

Surface plasmon resonance (SPR) is the resonant oscillation of conduction electrons at the interface between negative and positive permittivity material stimulated by incident light. SPR is the basis of many standard tools for measuring adsorption of material onto planar metal (typically gold or silver) surfaces or onto the surface of metal nanoparticles. It is the fundamental principle behind many color-based biosensor applications and different lab-on-a-chip sensors.

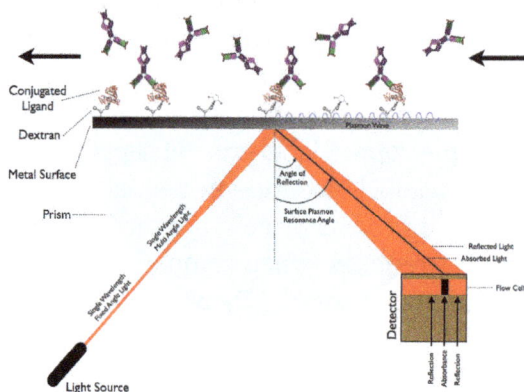

Surface plasmon resonance (SPR).

Explanation

The surface plasmon polariton is a non-radiative electromagnetic surface wave that propagates in a direction parallel to the negative permittivity/dielectric material interface. Since the wave is on the boundary of the conductor and the external medium (air, water or vacuum for example), these oscillations are very sensitive to any change of this boundary, such as the adsorption of molecules to the conducting surface.

To describe the existence and properties of surface plasmon polaritons, one can choose from various models (quantum theory, Drude model, etc.). The simplest way to approach the problem is to treat each material as a homogeneous continuum, described by a frequency-dependent relative permittivity between the external medium and the surface. This quantity, hereafter referred to as the materials' "dielectric function," is the complex permittivity. In order for the terms that describe the electronic surface plasmon to exist, the real part of the dielectric constant of the conductor must be negative and its magnitude must be greater than that of the dielectric. This condition is met in the infrared-visible wavelength region for air/metal and water/metal interfaces (where the real dielectric constant of a metal is negative and that of air or water is positive).

LSPRs (Localized SPRs) are collective electron charge oscillations in metallic nanoparticles that are excited by light. They exhibit enhanced near-field amplitude at the resonance wavelength. This field is highly localized at the nanoparticle and decays rapidly away from the nanoparticle/dieletric interface into the dielectric background, though far-field scattering by the particle is also enhanced by the resonance. Light intensity enhancement is a very important aspect of LSPRs and localization means the LSPR has very high spatial resolution (subwavelength), limited only by the size of nanoparticles. Because of the enhanced field amplitude, effects that depend on the amplitude such as magneto-optical effect are also enhanced by LSPRs.

Implementations

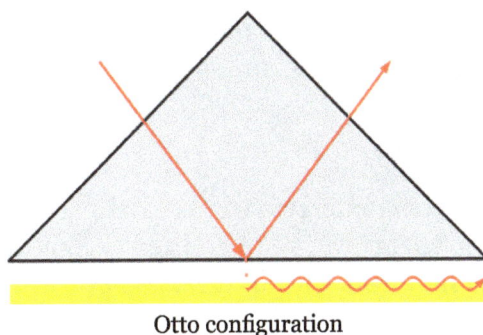

Otto configuration

In order to excite surface plasmons in a resonant manner, one can use electron bombardment or incident light beam (visible and infrared are typical). The incoming beam has to match its momentum to that of the plasmon. In the case of p-polarized light (polarization occurs parallel to the plane of incidence), this is possible by passing the light through a block of glass to increase the wavenumber (and the momentum), and achieve the resonance at a given wavelength and angle. S-polarized light (polarization occurs perpendicular to the plane of incidence) cannot excite electronic surface plasmons. Electronic and magnetic surface plasmons obey the following dispersion relation:

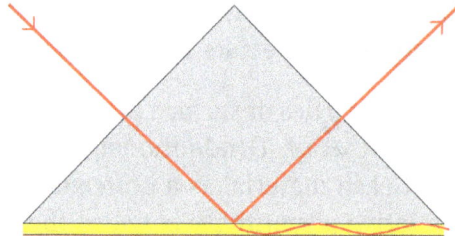

Kretschmann configuration

$$K(\omega) = \frac{\omega}{c}\sqrt{\frac{\varepsilon_1\varepsilon_2\mu_1\mu_2}{\varepsilon_1\mu_1 + \varepsilon_2\mu_2}}$$

where ϵ is the relative permittivity, and μ is the relative permeability of the material (1: the glass block, 2: the metal film).

Typical metals that support surface plasmons are silver and gold, but metals such as copper, titanium or chromium have also been used.

When using light to excite SP waves, there are two configurations which are well known. In the Otto setup, the light illuminates the wall of a glass block, typically a prism, and is totally internally reflected. A thin metal film (for example gold) is positioned close enough to the prism wall so that an evanescent wave can interact with the plasma waves on the surface and hence excite the plasmons.

In the Kretschmann configuration, the metal film is evaporated onto the glass block. The light again illuminates the glass block, and an evanescent wave penetrates through the metal film. The plasmons are excited at the outer side of the film. This configuration is used in most practical applications.

SPR Emission

When the surface plasmon wave interacts with a local particle or irregularity, such as a rough surface, part of the energy can be re-emitted as light. This emitted light can be detected *behind* the metal film from various directions.

Applications

Surface plasmons have been used to enhance the surface sensitivity of several spectroscopic measurements including fluorescence, Raman scattering, and second harmonic generation. However, in their simplest form, SPR reflectivity measurements can be used to detect molecular adsorption, such as polymers, DNA or proteins, etc. Technically, it is common to measure the minimum angle of reflection (maximum angle of absorption). This angle changes in the order of 0.1° during thin (about nm thickness) film adsorption. In other cases the changes in the absorption wavelength is followed. The mechanism of detection is based on that the adsorbing molecules cause changes in the local index of refraction, changing the resonance conditions of the surface plasmon waves. The same principle is exploited in the recently developed competitive platform based on loss-less dielectric multilayers (DBR), supporting surface electromagnetic waves with sharper resonances (Bloch surface waves).

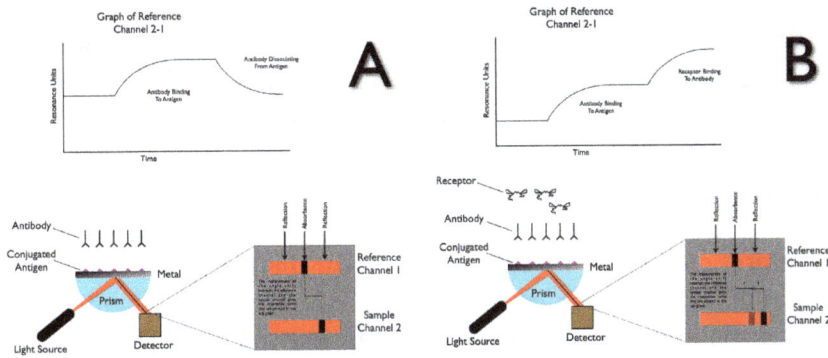

If the surface is patterned with different biopolymers, using adequate optics and imaging sensors (i.e. a camera), the technique can be extended to surface plasmon resonance imaging (SPRI). This method provides a high contrast of the images based on the adsorbed amount of molecules, somewhat similar to Brewster angle microscopy (this latter is most commonly used together with a Langmuir–Blodgett trough).

For nanoparticles, localized surface plasmon oscillations can give rise to the intense colors of suspensions or sols containing the nanoparticles. Nanoparticles or nanowires of noble metals exhibit strong absorption bands in the ultraviolet-visible light regime that are not present in the bulk metal. This extraordinary absorption increase has been exploited to increase light absorption in photovoltaic cells by depositing metal nanoparticles on the cell surface. The energy (color) of this absorption differs when the light is polarized along or perpendicular to the nanowire. Shifts in this resonance due to changes in the local index of refraction upon adsorption to the nanoparticles can also be used to detect biopolymers such as DNA or proteins. Related complementary techniques include plasmon waveguide resonance, QCM, extraordinary optical transmission, and dual polarization interferometry.

SPR Immunoassay

The first SPR immunoassay was proposed in 1983 by Liedberg, Nylander, and Lundström, then of the Linköping Institute of Technology (Sweden). They adsorbed human IgG onto a 600-angstrom silver film, and used the assay to detect anti-human IgG in water solution. Unlike many other immunoassays, such as ELISA, an SPR immuno-assay is *label free* in that a label molecule is not required for detection of the analyte. Additionally, the measurements on SPR can be followed real-time allowing the monitoring of individual steps in sequential binding events particularly useful in the assessment of for instance sandwich complexes.

Material Characterization

Multi-Parametric Surface Plasmon Resonance, a special configuration of SPR, can be used to characterize layers and stacks of layers. Besides binding kinetics, MP-SPR can also provide information on structural changes in terms of layer true thickness and refractive index. MP-SPR has been applied successfully in measurements of lipid targeting and rupture, CVD-deposited single monolayer of graphene (3.7Å) as well as micrometer thick polymers.

Data Interpretation

The most common data interpretation is based on the Fresnel formulas, which treat the formed thin films as infinite, continuous dielectric layers. This interpretation may result in multiple possible refractive index and thickness values. However, usually only one solution is within the reasonable data range. In Multi-Parametric Surface Plasmon Resonance, two SPR curves are acquired by scanning a range of angles at two different wavelengths, which results in a unique solution for both thickness and refractive index.

Metal particle plasmons are usually modeled using the Mie scattering theory.

In many cases no detailed models are applied, but the sensors are calibrated for the specific application, and used with interpolation within the calibration curve.

Examples

Layer-by-layer Self-assembly

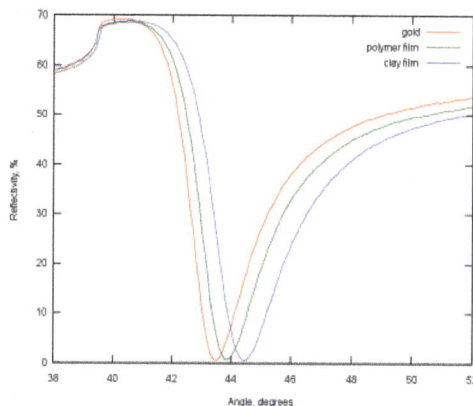

SPR curves measured during the adsorption of a polyelectrolyte and then a clay mineral self-assembled film onto a thin (ca. 38 nanometers) gold sensor.

One of the first common applications of surface plasmon resonance spectroscopy was the measurement of the thickness (and refractive index) of adsorbed self-assembled nanofilms on gold substrates. The resonance curves shift to higher angles as the thickness of the adsorbed film increases. This example is a 'static SPR' measurement.

When higher speed observation is desired, one can select an angle right below the resonance point (the angle of minimum reflectance), and measure the reflectivity changes at that point. This is the so-called 'dynamic SPR' measurement. The interpretation of the data assumes that the structure of the film does not change significantly during the measurement.

Binding Constant Determination

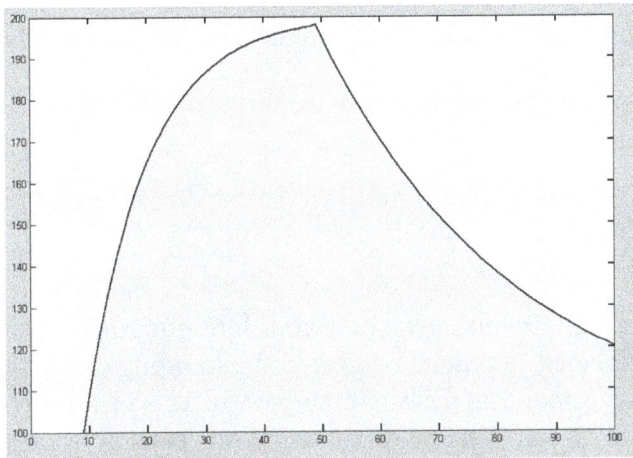

Association and dissociation signal

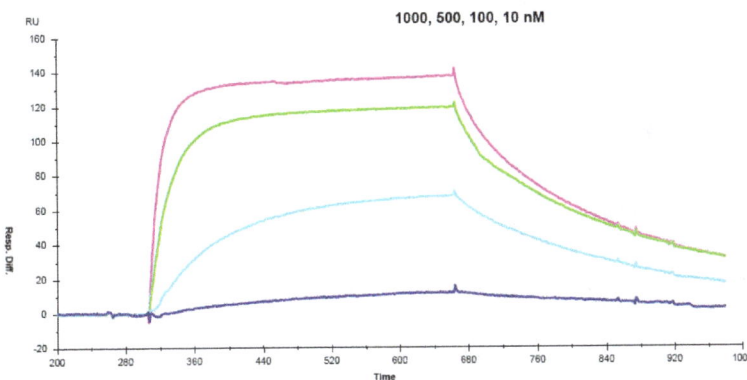

Example of output from Biacore

When the affinity of two ligands has to be determined, the binding constant must be determined. It is the equilibrium value for the product quotient. This value can also be found using the dynamic SPR parameters and, as in any chemical reaction, it is the association rate divided by the dissociation rate.

For this, a bait ligand is immobilized on the dextran surface of the SPR crystal. Through a microflow system, a solution with the prey analyte is injected over the bait layer. As the prey analyte binds the bait ligand, an increase in SPR signal (expressed in response units, RU) is observed. After desired association time, a solution without the prey analyte (usually the buffer) is injected on the microfluidics that dissociates the bound complex between bait ligand and prey analyte. Now as the prey analyte dissociates from the bait ligand, a decrease in SPR signal (expressed in resonance units, RU) is observed. From these association ('on rate', k_a) and dissociation rates ('off rate', k_d), the equilibrium dissociation constant ('binding constant', K_D) can be calculated.

The actual SPR signal can be explained by the electromagnetic 'coupling' of the incident light with the surface plasmon of the gold layer. This plasmon can be influenced by the layer just a few nanometer across the gold–solution interface i.e. the bait protein and possibly the prey protein. Binding makes the reflection angle change;

$$K_D = \frac{k_d}{k_a}$$

Thermodynamic Analysis

As SPR biosensors facilitate measurements at different temperatures, thermodynamic analysis can be performed to obtain a better understanding of the studied interaction. By performing measurements at different temperatures, typically between 4 and 40 °C, it is possible to relate association and dissociation rate constants with activation energy and thereby obtain thermodynamic parameters including binding enthalpy, binding entropy, Gibbs free energy and heat capacity.

Pair-wise Epitope Mapping

As SPR allows real-time monitoring, individual steps in sequential binding events can be thoroughly assessed when investigating the suitability between antibodies in a sandwich configuration. Additionally, it allows the mapping of epitopes as antibodies of overlapping epitopes will be associated with an attenuated signal compared to those capable of interacting simultaneously.

Magnetic Plasmon Resonance

Recently, there has been an interest in magnetic surface plasmons. These require materials with large negative magnetic permeability, a property that has only recently been made available with the construction of metamaterials.

I. Basic Working Principle of Surface Plasmon Resonance

SPR happens when energy from monochromatic incident light beam hits the metal-dielectric interface at a particular SPR angle, gets transformed into electromagnetic energy resulting into production of evanescent waves.

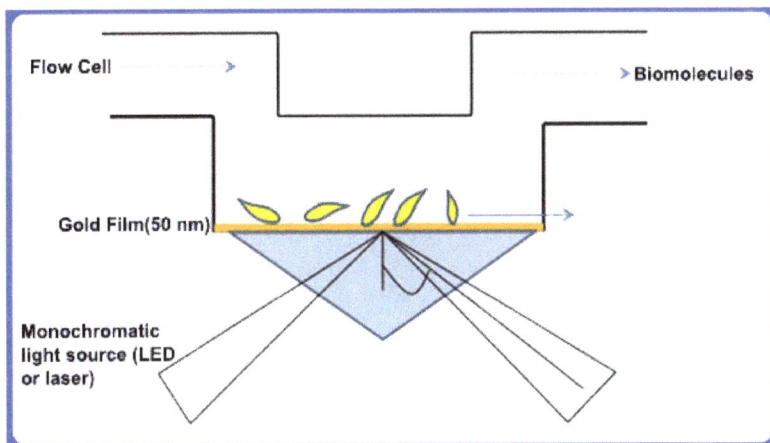

The basic principle of SPR: Measurement of alterations in refractive
index of medium directly in contact with sensor surface.

Gold surface is generally used in SPR for immobilization of test proteins. The unlabelled query molecules are introduced in solution form and alterations in angle of reflection of light due to the binding of the probes to the immobilized protein provides real-time information regarding biomolecular interactions.

The angle at which the minimum intensity of the reflected light is achieved is called "SPR angle". SPR angle is directly related to the quantity of biomolecules bound to the sensor surface. Different factors such as nature of the metal layer, angle of SPR, refractive index at the metal-dielectric interface, wavelength of the incident light, etc regulate the magnitude of surface plasmon resonance.

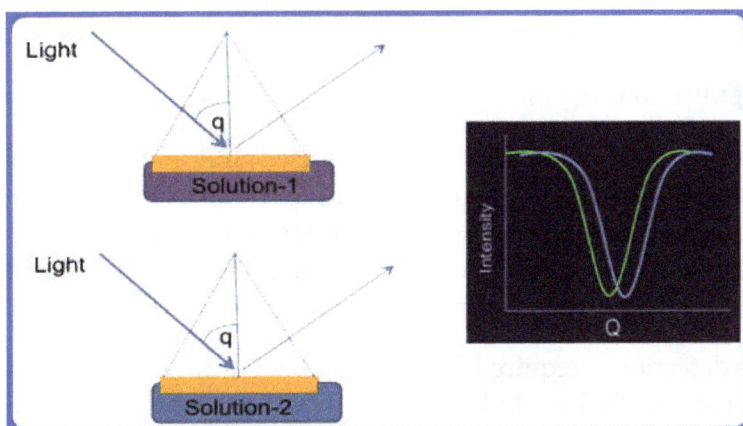

SPR angle; effect of refractive index near the surface on SPR angle

In SPR measurement the sensorgram indicates the changes in reflection intensity with respect to incident angle before and after binding of the target molecule.

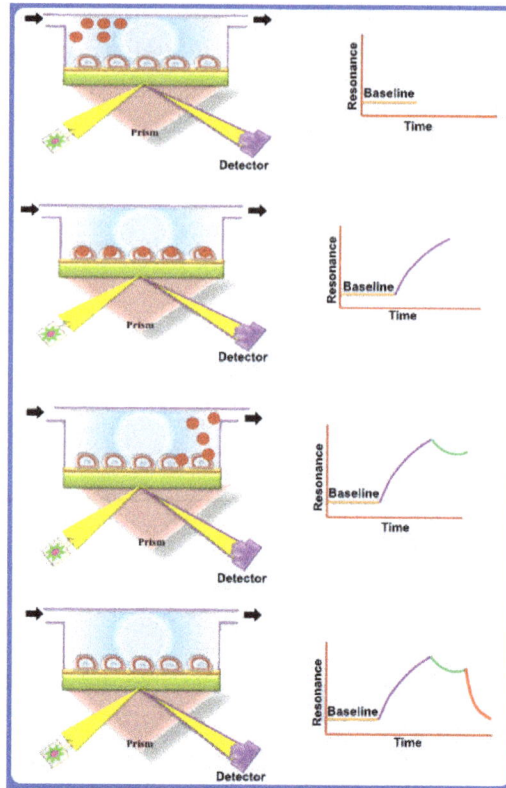

Different steps involved in biomolecular interaction analysis using SPR sensor. Overall SPR-based biosensors contain the incident light source, sensor gold surface, and the detector to capture the reflected light. The interaction between the molecules is measured by plotting the sensorgram (as shown in panels from top to bottom). The ligand molecules are immobilized on an activated gold sensor chip; while the query molecule in buffer is flowed through the flow cell. Interaction along with binding kinetics is monitored in a real-time manner.

II. Different Applications of Spr

In last ten years, several research groups have used SPR and related label-free techniques for real-time analysis of protein-protein and other biomolecular interactions, measurement of low abundance serum biomarkers, and screening of inhibitors of tumor targets and potential drug molecules. SPR has also been applied extensively for many biomedical, food and environmental applications.

Ultra-sensitive detection is required for measurement of very low-abundance biomarkers. In a recent study Choi et al., have utilized SPR-based biosensor for detection of prostate-specific antigen (PSA) (Choi et al., 2008). In this study the authors have employed gold (Au) nanoparticle–antibody complex for signal enhancement of SPR and thereby effectively increased the sensitivity of the detection approach. SPR-based immunosensor have been designed, where a gold surface coated with PSA monoclonal antibodies (mAbs) and gold nanoparticle–conjugated antibody complex was used to capture PSA antigen. With this SPR-based biosensor the authors were able to detect PSA with a detection limit of 300 fM.

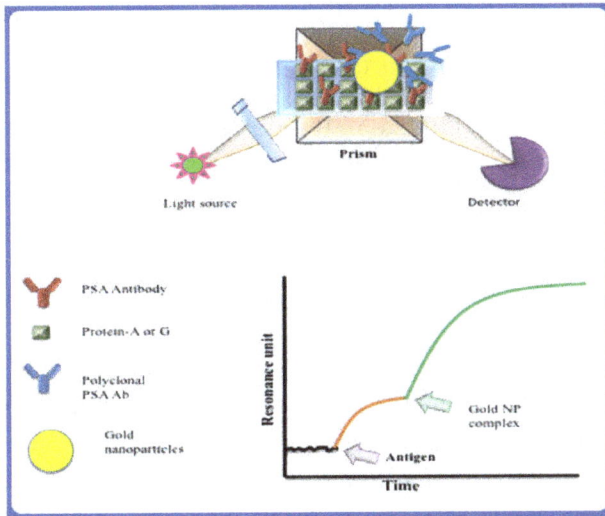

Sandwich immunoassay SPR platform for detection of cancer biomarker; prostate specific antigen (PSA) PSA monoclonal antibodies were coated on a gold sensor chip functionalized with recombinant protein G via thiol groups. The sensorgram indicates the binding of the target antigen with immobilized antibodies. However, the change in SPR angle was not enough to monitor such interactions in complex clinical samples. A sandwich immunoassay approach using gold nanoparticle conjugated with PSA polyclonal antibody complex amplified the signal. Use of gold nanoparticle-antibody complex as signal amplifier significantly improved the sensitivity level.

III. Advantages and Disadvantages

SPR has its own advantages and limitations like other technologies.

Advantages

- Label-free detection eliminates the necessity of any secondary reactants and lengthy labeling process.

- Real-time measurements of biomolecular interaction.

- Multiplex analysis is possible; therefore, compatible with high-throughput assays.

- Sensitive to conformational changes; direct measurements and study of binding kinetics are possible.

- Both quantitative and qualitative measurements are possible.

Disadvantages

- Requires sophisticated instrumentation.

- Restricted to choice of metal (gold/silver surfaces).

- Completely dependent on mass changes.

- Temperature sensitive; change in temperature affects refractive index.

- Non-specific interactions affect the signal.

- Works efficiently only with homogeneous surface.

- Sensitivity and specificity often become major concerns while handling very complex biological samples.

It is evident that SPR-based label-free sensors are very promising for detection of disease biomarkers and HT real-time screening of biomolecules. However, SPR and related technologies have limited usage for large-scale applications in industries and clinics due to multiple technical limitations discussed above. Current advancements in the field of SPR have introduced quite a few new materials and methods for the improvement of sensitivity of the instruments. Application of different polarization methods for the incident light (such as p-polarised, s-polarised, TM waves, TE waves) can effectively enhance the coupling of incident light to plasmons. In addition, there are constant efforts to make reproducible and well-established surface chemistry for the generation of a selective sensing interface, which can reduce the binding of non-specific moieties that can alter the SPR signal. Multi-dimensional applications of SPR-based techniques have been achieved through successful coupling with different technological approaches, including SPR-MS, electrochemical SPR and surface plasmon fluorescence spectroscopy, which effectively circumvented some of the basic limitations associated with SPR. These efforts can expand the applications of SPR-based biosensors in clinical diagnosis.

Surface Plasmon Resonance Imaging

SPRi-based label-free sensors, which depend on measurement of an inherent property (refractive index) of query molecules, are gaining popularity in HT proteomics due to their multiplexing capabilities. The basic working principle of SPRi is quite similar to that of SPR, which measures the changes in the refractive index. In SPR imaging (SPRi) a spatially resolved imaging device is introduced to SPR set-up for continuously monitoring the changes occuring on the surface to generate real-time kinetic data. SPRi-based biosensors are capable for instantaneous label-free analysis of several biomolecular interactions in a fast and HT manner (Reddy et al., 2012).

I. Basic Working Principle of Surface Plasmon Resonance Imaging (SPRi)

In SPRi the alterations in the refractive index of the medium directly in contact with sensor surface is measured.

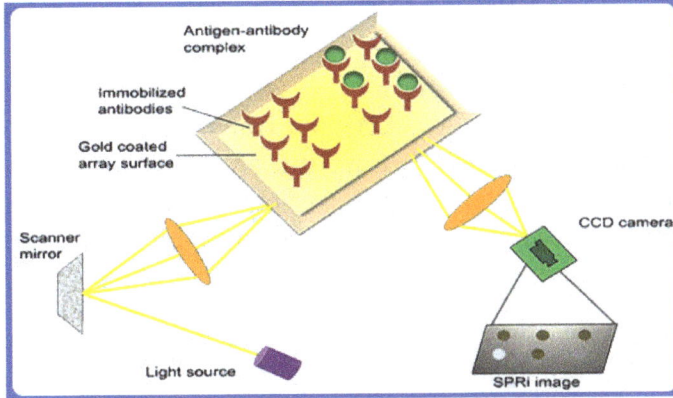

SPRi relies on measurement of changes in refractive
index of the medium directly in contact with sensor surface.

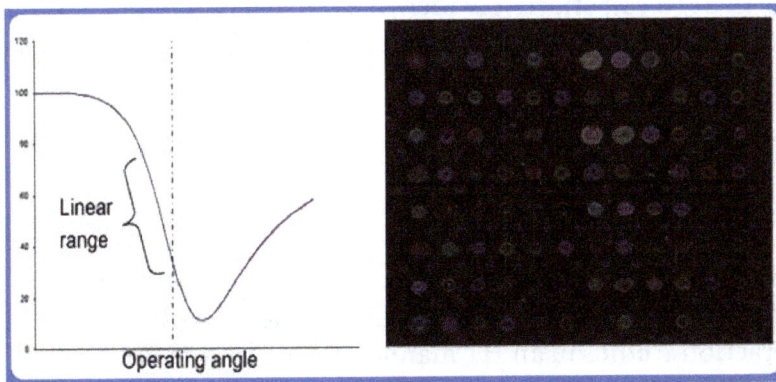

Selection of an operating angle and assignment of region of interests (ROIs) in SPRi experiment.

In SPRi complete biochip surface is illuminated using a broad beam, monochromatic, polarized light and a CCD camera is used for simultaneous capturing of reflected light from each spot. In SPRi the intensity of the incident light as well as wavelength are kept constant, and the reflected light is determined at an optimum reflectance angle coming from the metal interface. The complete array can be captured by coupled charge device (CCD) camera for HT studies. SPRi experimental flow includes preparation and mounting of the slide, loading prime samples, assignment of region of interests (ROIs), determination of operating angle and data acquisition.

In SPRi experiments data generation is done in a real-time manner and generated data files are saved in proper format (as movies) and exported for further analysis.

II. Different Applications of SPRi

SPRi-based biosensors are attractive choices for detection of low-abundance analytes and biomolecular interactions in HT manner (Ray et al., 2010; Reddy et al., 2012). Over the last decade quite a few studies have used this label-free sensing platform for direct detection of potential maker proteins in human serum and other biological samples. In a recent study, Ladd et al. has applied SPRi in combination with antibody array for detection of cancer biomarkers in buffer solution and diluted human serum (10%) samples (Ladd et al., 2009). The authors have employed SPR imaging sensor with polarization contrast and detected activated leukocyte cell adhesion molecule/CD 166 (ALCAM) and transgelin-2 (TAGLN2) biomarkers, down to ng/mL concentration (LOD 6 ng/mL and 3 ng/mL for ALCAM and TAGLN2, respectively) devoid of any significant cross-reactivity. Such studies testify the potential of SPRi-based biosensors for detection of biomarkers from complex biological samples.

However, the sensitivity obtained with actual biological samples are found to be less compared to that obtained with buffer solutions. To increase the sensitivity of SPRi-based biosensors different nanostructured materials, particularly quantum dots and goldnanoparticles are introduced for signal amplification. Application of nanoparticles in SPRi can effectively increase the detection limit and target antigen can be detected at pM level, which is not possible to achieve with conventional SPRi settings

III. Advantages and Disadvantages of SPRi

Advantages

The major advantage of SPRi over standard SPR is its multiplexing capabilities (ability to observe hundreds of reactions simultaneously). SPRi can be applied to perform interaction studies in an HT manner, which is not possible by conventional SPR. Additionally; other advantages of SPR-based biosensors are also applicable for SPRi.

- Label-free detection eliminates the requirement of any secondary reactants and long labeling process.

- Real-time measurements of biomolecular interaction and binding kinetics.

- Highly potential for multiplex analysis; so very well suited for high-throughput analysis.

- Both quantitative and qualitative measurements are possible.

Disadvantages

Due to its HT capabilities, SPRi is promising for clinical research, but there are many limitations as well; such as

- Requires sophisticated instrumentation.

- Restricted to choice of metal (gold/silver surfaces).

- Changes in temperature affects refractive index and thereby efficiency of the measurement.

- Non-specific interactions affect the signal.

- Heterogeneous sample surface affects sensitivity.

Not very effective in handling complex biological samples.

SPRi-based biosensors have shown their potential to measure kinetic reactions of bio-molecular interactions in HT manner. These techniques have been found to be very efficient for real-time measurement of disease-related proteins in buffer in vitro. Al-though, some of the recent studies have testified the applicability of SPRi-based bio-sensors for direct detection of marker proteins in different biological fluids, including serum/plasma, saliva and urine. Issues regarding sensitivity and specificity remain to be explored further when complex biological samples are concerned. Due to the re-quirement of sophisticated instrumentation and restriction to gold/silver surfaces, the detection cost of SPRi-based sensors is very high, and not affordable for routine clinical diagnostics. If these basic limitations are circumvented successfully; SPRi could be one of the very attractive choice for HT proteomic research.

I. Surface Plasmon Resonance (SPR) and Spr Imaging (SPRi); Comparative Analysis

SPR-based sensing approaches depend on measurement of changes in the refractive index of medium directly in contact with sensor surface. In SPR measurement the sensorgram indicates the changes in reflection intensity with respect to incident angle when the target molecule binds to the sensor surface. Alteration in the refractive angle is directly proportional to the mass bound at the surface. Although SPR and SPRi fol-low the same working principle for bio-sensing, there are following major differences in SPRi instrumental set-ups:

- In SPRi a CCD camera is used for instantaneous capturing of reflected light from each spot of sensor surface, which allows instantaneous analysis of bind-ing events on all spots.

- SPRi is more suitable for HT analysis compared to SPR

Both SPR and SPRi have the following advantages regarding analysis of protein-pro-tein interactions:

- Allow label-free detection without need of any secondary reactants and exten-sive labeling process

- Real-time measurements

- Provide information regarding binding kinetics (rates of association and disso-ciation)

- Suitable for studying multiple biomolecular interactions simultaneously

- Provide both quantitative and qualitative measurements

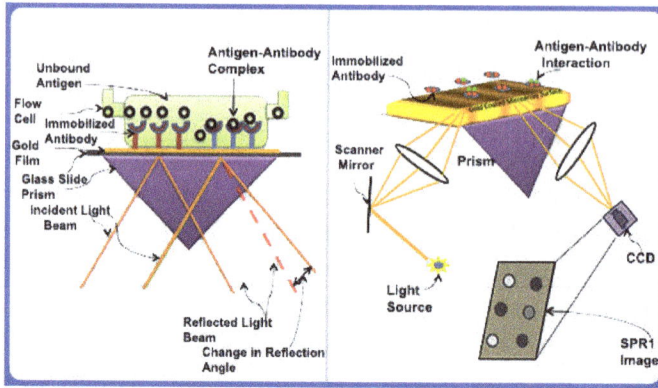

The basic working principle of SPR (A) and SPRi (B) for monitoring protein-protein interactions. In this label-free detection method one of the interaction partners is immobilized on the sensor surface and the second protein is allowed to pass-through. Interaction between the interacting protein partners are monitored by measuring changes in the refractive angle, which is inversely proportional to the mass bound at the sensor surface.

II. Study of Protein-Protein Interactions using SPR and SPRi

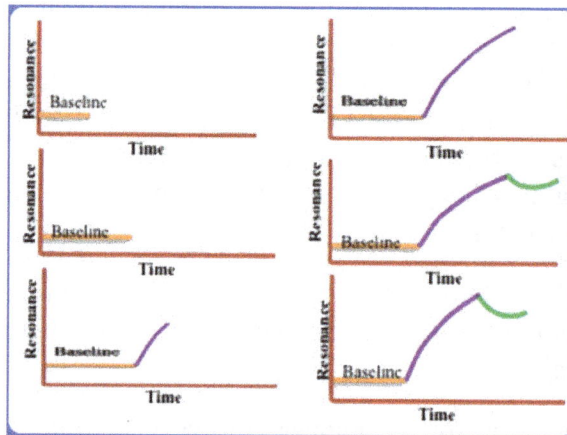

Determination of association and dissociation kinetics of biomolecular interaction using SPR. Different steps involved in biomolecular interaction analysis using SPR sensor. The interaction between the molecules is measured by plotting the sensorgram as shown in panels.

SPR-based sensing approaches have been successfully used in studying protein-protein and other biomolecular interactions. In order to analyze interactions between two proteins using SPR-based sensing approaches; one of the interaction partners is immobilized on the sensor surface, while the other one is injected in solution (Willander and Al-Hilli 2009). Binding of second partner on the sensor surface due to biomolecular interactions leads to changes in the refractive angle, which is inversely proportional to the mass bound at the sensor surface. The quantity of bound materials is continuously measured as a function of time. After monitoring the interactions for certain period of time, the buffer solution is changed to stop the interaction to monitor the dissociation of the complex formation.

SPR provides real-time measurements of interactions; for that reason kinetics of interactions [on rate (Ka) and off rate (Kd)] can be calculated.

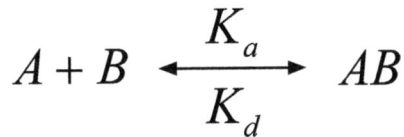

$$A + B \xleftrightarrow[K_d]{K_a} AB$$

Association rate constant of interaction can be calculated from the association period data, if the concentration of interacting partners is known. To measure the dissociation rate constant(s) exponential decay to the dissociation data are used (Berggård et al., 2007).

There are several published studies where SPR/SPRi has been used for measuring protein-protein or protein-small molecule interactions; few studies are described in-tabulated formats.

Application of SPR and SPRi for studying protein-protein or protein-small molecule interactions

Study	Detection technique	Description
Antigen-antibody (IL6 and anti-IL6) interactions (Rispens et al., 2011)	SPR	Interaction between anti-interleukin 6 (IL6) antibodies, anti-IL6.16 and anti-IL6.8 with interleukin 6 (chips coated with either anti-IL6.16 or anti-IL6.8). Time point analysis was performed using different concentrations of IL6 and two antibodies (anti-IL6.16 and anti-IL6.8). Higher on-rate and smaller off-rate was found for anti-IL6.8 than anti-IL6.16, indicating higher affinity of anti-IL6.8 for interleukin 6.
Studying association or dissociation kinetics of diverse Fab and hK1 interaction (Wassaf et al., 2006)	SPR in combination with microarray	The overall aim of this study was detection of antibodies that bind at the active site of human tissue kallikrein 1 (hK1) and consequently inhibit the protease activity of hK1. Simultaneous analysis of kinetic constants (kon and koff) for 96 different Fab fragments using array format. Fabs were categorized on basis of their capacity to recognize an apparent active site epitope. Immobilization of Fab was performed using specific capture surfaces (anti-cMyc or protein A).
IgG and Protein A interaction (Natarajan et al., 2008)	SPRi in combination with continuous-flow microfluidics (CFM)	This study demonstrated that coupling of continuous-flow microfluidics with SPRi could improve printing process. It allows immobilization of purified proteins on gold surface at a very low concentration. Low sample consumption and multiplexing capability are major advantages of this combined technology.

Clinically related protein-peptide interactions (Cherif et al., 2006)	SPRi	Interactions of three peptides (C 20–40, C 131–150, and Ova 273–288) with rabbit anti-C 20–40 and anti-C 131–150 immune sera (1:100 diluted). Detection of such weak clinically relevant interactions is promising for clinical research.
Antibody-antigen binding (Nogues et al., 2010)	SPRi in combination with peptide microarrays	65kDa isoform of human glutamate decarboxylase (GAD65) and a human monoclonal antibody. Specific bindings of Rac1 and RhoA antibodies to their antigens immobilized on protein arrays were monitored by spectral SPR imaging. Kinetic parameters of the interaction were measured more efficiently than ELISA/RIA methods.
Interactions of GST-fusion proteins with their antibodies (Yuk et al., 2006)	SPRi (Wavelength interrogation-based self-constructed with Kretschmann-Raether geometry)	Interactions of glutathione S-transferase-fusion proteins with their antibodies; anti-Rac1 and anti-RhoA to Rac1 and RhoA have been studied. Quartz tungsten halogen lamp was applied as a light source. Protein arrays were prepared by immobilizing glutathione S-transferase (GST) fusion proteins on the glutathione surfaces. Protein arrays were analyzed by two-dimensional images.

III. Data Processing and analysis

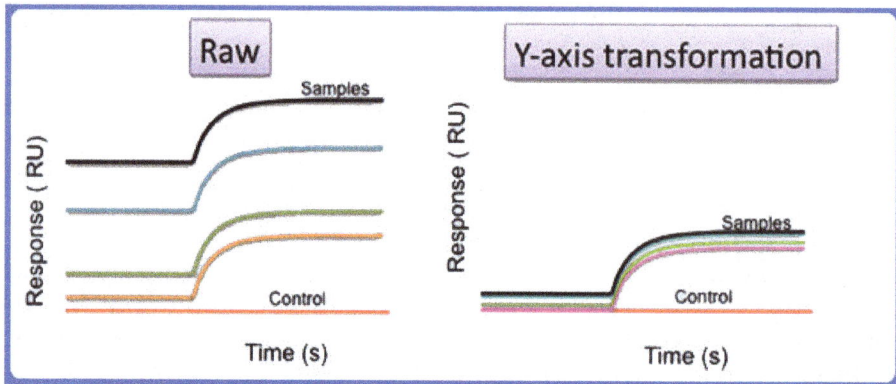

Y-axis transformation of SPR sensorgram to fit data

Processing of SPR raw data prior to analysis is essential to obtain superior interpretation of data generated during the association/ dissociation phase of SPR-based analysis of protein-protein interactions. After obtaining raw-data from experimental process Y-axis transformation is performed to fit the baselines of different time points and concentration variations. There are different commercially available software to fit curves and obtain apparent rate constants. More often presence of background spot intensity significantly reduces artifacts and improves the S/N ratio, and thereby quality of the data. To obtain sensorgram for kinetics analysis subtraction of background spots are essential. Response from reference surface and response of buffer injection are subtracted before analysis of SPR datasets. Considering a bidirectional interaction between two interacting partners the value for KD (ratio of Kon/Koff) is generally calculated using a nonlinear least-squares analysis (O'Shannessy et al., 1993 & 1994).

To obtain kinetic constants various binding models, like simple Langmuir binding models are applied. Controls are prepared using empty surface for baseline checking. While determination of kinetics of reactions between two interacting partners, one partner is immobilized and multiple concentrations of second analyte are used and interactions are analyzed. Quality of the obtained data can be evaluated by verifying the Kon and Koff values.

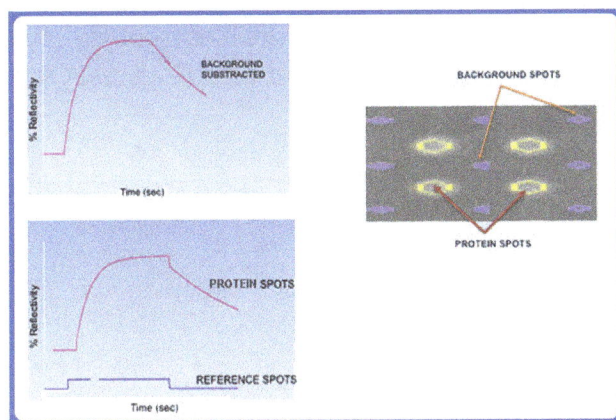

Subtraction of background reference spot intensity

Kinetics of binding is very crucial while selection of drug targets and screening of potential drug molecules. In SPR and SPRi the major advantage is that multiple interactions can be monitored simultaneously, which allows detection of interactions of two interacting partners using different concentration of ligands. The interaction of multiple analytes to same ligand can also be monitored and binding affinities can be compared. Even if multiple molecules have same affinity, they may have different Kon and Koff, and specific candidates can be selected according to the need of the investigator.

Although SPR and SPRi-based biosensors are suitable for diverse types of applications, maximum use of those label-free sensing approaches are found to be in analysis of biomolecular interactions, since it provides real-time information regarding equilibrium binding constants, kinetic rate constants and thermodynamic parameters. Over the last decade large number of studies have proved the utility of SPR and SPRi-based biosensors for protein-protein and protein-small molecule interactions. Additionally, interactions of proteins with other types of biomolecules including protein-carbohydrate interactions DNA-protein interaction, etc. are also vividly studied by SPR and SPRi. In spite of multiple advantages, SPR-based approaches have not yet gained extreme popularity in routine clinical use, mostly due to the detection cost associated with requirement of sophisticated instrumentation and restriction to gold/silver surfaces. To extend the applications of SPR-based techniques, now diverse amalgamated technological approaches have been developed such as SPR-MS, electrochemical SPR and surface plasmon fluorescence spectroscopy, which have successfully circumvented some of the basic limitations associated with SPR.

Bio-layer Interferometry

Bio-layer interferometry (BLI) is a label-free technology for measuring biomolecular interactions within the interactome. It is an optical analytical technique that analyzes the interference pattern of white light reflected from two surfaces: a layer of immobilized protein on the biosensor tip, and an internal reference layer (Figure A). Any change in the number of molecules bound to the biosensor tip causes a shift in the interference pattern that can be measured in real-time (Figures A and B).

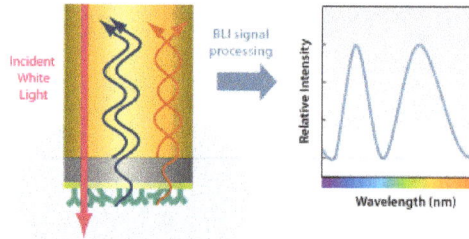

Figure A

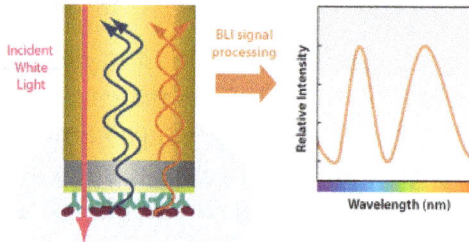

Figure B

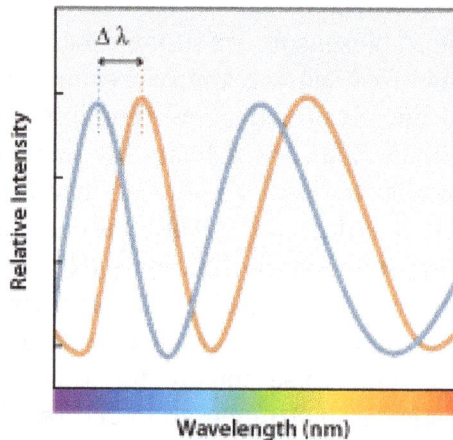

Figure C

The binding between a ligand immobilized on the biosensor tip surface and an analyte in solution produces an increase in optical thickness at the biosensor tip, which results in a wavelength shift, $\Delta\lambda$ (Figure C), which is a direct measure of the change in thickness of the biological layer. Interactions are measured in real time, providing the ability to monitor binding specificity, rates of association and dissociation, or concentration, with precision and accuracy.

Only molecules binding to or dissociating from the biosensor can shift the interference pattern and generate a response profile. Unbound molecules, changes in the refractive index of the surrounding medium, or changes in flow rate do not affect the interference pattern. This is a unique characteristic of bio-layer interferometry and extends its capability to perform in crude samples used in applications for protein-protein interactions, quantitation, affinity, and kinetics.

Bio-layer interferometry was pioneered by the founders of ForteBio, an instrument manufacturer with headquarters in Menlo Park California.

References

- Shizuya, H; Kouros-Mehr, H (2001). "The development and applications of the bacterial artificial chromosome cloning system". The Keio journal of medicine. 50 (1): 26–30. PMID 11296661. doi:10.2302/kjm.50.26

- Steinfeld J.I., Francisco J.S. and Hase W.L. Chemical Kinetics and Dynamics (2nd ed., Prentice-Hall 1998) p.263 ISBN 0-13-737123-3

- "Investigating luminol" (PDF). Salters Advanced Chemistry. Archived from the original (PDF) on September 20, 2004. Retrieved 2006-03-29

- Hupf, H.B.; Eldridge, J.S.; Beaver, J.E. (1968). "Production of Iodine-123 for medical applications". The International Journal of Applied Radiation and Isotopes. 19 (4): 345–346. PMID 5650883. doi:10.1016/0020-708X(68)90178-6

- Ashutosh Sharma; Stephen G. Schulman (21 May 1999). Introduction to Fluorescence Spectroscopy. Wiley. ISBN 978-0-471-11098-9

- Gilby, ED; Jeffcoate, Edwards (July 1973). "125-Iodine tracers for steroid radioimmunoassay.". Journal of Endocrinology. 58 (1): xx. PMID 4578967

- Radiation Protection and the Management of Radioactive Waste in the Oil and Gas Industry (PDF) (Report). International Atomic Energy Agency. 2003. pp. 39–40. Retrieved 20 May 2012

- Rauhut, Michael M. (1985), Chemiluminescence. In Grayson, Martin (Ed) (1985). Kirk-Othmer Concise Encyclopedia of Chemical Technology (3rd ed), pp 247 John Wiley and Sons. ISBN 0-471-51700-3

- Human Framework Adaptation of a Mouse Anti-Human IL-13 Antibody Fransson, J. and others. Journal of Molecular Biology, 2010, 398 (2), 214-231

Permissions

Index